Using Excel

For Statistics and Data Analysis

Glenn Zabowski

McGill University

Addison-Wesley

An imprint of Addison Wesley Longman Ltd.

Don Mills, Ontario • Reading, Massachusetts • Harlow, England
Melbourne, Australia • Amsterdam, The Netherlands • Bonn, Germany

Publisher: Ron Doleman
Managing Editor: Linda Scott
Editor: Suzanne Schaan
Cover Design, Design: Anthony Leung
Desktopping: Maria Modopoulos/Design Tech 2000
Production Coordinator: Alexandra Odulak
Manufacturing Coordinators: Jodi Carbert, Sharon Latta Paterson

Microsoft is a registered trademark of Microsoft Corporation.

Canadian Cataloguing in Publication Data

Zabowski, Glenn
  Using Excel for statistics and data analysis

Includes index.
ISBN 0-201-39621-1

1. Microsoft Excel (Computer file).   2. Electronic spreadsheets—Computer programs.   3. Business—Computer programs.   I. Title.

HF5548.4.M523Z32 1999   650'.0285'5369   C98-933-062-1

Copyright © 1999 Addison Wesley Longman Ltd.
All rights reserved. No part of this publication may be reproduced, or stored in a database or retrieval system, distributed, or transmitted in any form or by any means, electronic, mechanical, photocopying, recording or otherwise, without the prior written permission of the publisher.

ISBN 0-201-39621-1

Printed and bound in Canada.

A B C D E -WC- 03 02 01 00 99

Contents

Preface v

Introduction: Excel Basics vii

Chapter 1: Presenting Data in Basic Charts
    1.1 Creating Pie Charts 1
    1.2 Creating Bar Charts and Column Charts 7
    1.3 Creating Line Charts and Frequency Polygons 11
    1.4 Creating Scatter (*xy*) Charts 17

Chapter 2: Describing Numerical Data
    2.1 Sorting Data into Ordered Arrays and Frequency Distributions 24
    2.2 Using Excel Functions to Produce Descriptive Statistics 28
    2.3 Using the Data Analysis Tool to Produce Descriptive Statistics 30
    2.4 Interpreting Descriptive Statistics 33
    2.5 Creating Histograms 37

Chapter 3: Tables for Categorical Data
    3.1 Creating One-Way Summary Tables for Categorical Variables 42
    3.2 Creating Two-Way Summary Tables for Categorical Variables 45

Chapter 4: Probability Distributions
    4.1 Computing Expected Values, Variances, and Standard Deviations 49
    4.2 Computing Binomial Probabilities 52
    4.3 Computing Hypergeometric Probabilities 56
    4.4 Computing Poisson Probabilities 59
    4.5 Computing Normal Probabilities 63
    4.6 Computing Exponential Probabilities 67

Chapter 5: Sampling and Simulating Sampling Distributions
    5.1 Selecting a Random Sample from a Population 70
    5.2 Using Random Numbers to Simulate the Central Limit Theorem 73

Chapter 6: Estimation
    6.1 Confidence Interval for the Population Mean Using z 77
    6.2 Confidence Interval for the Population Mean Using t 81
    6.3 Confidence Interval for the Population Proportion 85
    6.4 Determining the Sample Size for the Population Mean 87
    6.5 Determining the Sample Size for the Population Proportion 90

Chapter 7: Hypothesis Testing: One Population
    7.1   Creating a Hypothesis Test for One Population Mean (z-test) 93
    7.2   Creating a Hypothesis Test for One Population Mean (t-test) 97
    7.3   Creating a Hypothesis Test for One Population Proportion 101

Chapter 8: Hypothesis Testing: Two Populations
    8.1   Pooled Variance t-Test and Confidence Interval 105
    8.2   Matched Samples t-Test and Confidence Interval 112
    8.3 z-Test for Differences in Two Proportions 119
    8.4 F-Test for Two Population Variances 123

Chapter 9: Chi-Square Tests
    9.1   Chi-Square Test for Normality 128
    9.2   Contingency Table Test for Independence 132

Chapter 10: Analysis of Variance
    10.1   One-Way ANOVA F-Test 138

Chapter 11: Simple Linear Regression
    11.1   Inserting a Trend Line and the TREND Function 144
    11.2   Using the Data Analysis Tool for Simple Regression 148
    11.3   Residual Analysis 154
    11.4   Obtaining Confidence Intervals for $E(y)$ and Prediction
            Intervals for Individual Responses 157

Chapter 12: Multiple Regression
    12.1   Using the Data Analysis Tool for Multiple Regression 161
    12.2   Using the Data Analysis Tool for Correlation 168
    12.3   Indicator Variables 172

Chapter 13: Time-Series Analysis and Forecasting
    13.1   Moving Averages 177
    13.2   Exponential Smoothing 181
    13.3   Trend Fitting 184
    13.4   Classical Multiplicative Decomposition 185
    13.5   Regression Approach to Time Series 191

Chapter 14: Quality Control Charts
    14.1 $\bar{x}$ and R Charts 194
    14.2 p and c Charts 199

Index 205

Preface

Spreadsheet analysis has become an indispensable tool in business analysis. Excel is the premier spreadsheet program in today's business world, and students who know Excel have a definite edge in the job market. Many students ask for Excel in their courses, as they already realize the importance of spreadsheet skills in their future jobs. *Using Excel for Statistics and Data Analysis* is intended to be a supplementary workbook for introductory statistics courses; its main purpose is to demonstrate Excel's data analysis capabilities and to show how Excel can be used as a tool for statistical analysis. Although Excel is not a statistical software package, it can perform most of the data analysis required for an introductory business or economics statistics course, either undergraduate or MBA. Excel may not be appropriate for a more advanced course, where a full statistical package may be required for more advanced analysis.

Students using this workbook need not have any previous experience with Excel, although many will have a basic working knowledge of this software. Using Excel for statistical analysis will show students how statistics is related to other courses such as accounting, finance, and management information systems, since Excel is increasingly being used in those courses as well. Excel is more accessible than any statistical software, allowing students to feel more comfortable working with spreadsheets and to see more easily how statistics is relevant in making everyday business decisions.

FEATURES

Each chapter in *Using Excel for Statistics and Data Analysis* deals with a particular type of statistical analysis. Within the chapters, each section is devoted to a single procedure. The workbook's features include the following:

- **Statistics review:** Most sections begin with a brief outline of the relevant statistical theory and concepts, without actually teaching or developing them. Terms are defined, key formulas are given, and notation is specified. Students should refer to their core statistics text for further information.

- **Example:** Each section presents an example based on real-world data. The example illustrates how the particular analysis can be conducted either by using one of Excel's built-in data analysis tools or by inputting Excel formulas to set up templates.

- **"Setting Up the Template"**: Simple step-by-step instructions are included so that students can follow along by working through the example themselves on the computer. Screen grabs illustrate the use of Excel's Wizards, and formulas are provided for creating templates. While the instructions are based on Excel 97, most of them can be used with or adapted for other versions of the software.

- **"Discussing the Outcome"**: Once the spreadsheet has been constructed, the results are discussed and interpreted, and conclusions are drawn based on the output. Excel's spreadsheet capabilities are then used to observe the effect of changes in data values on statistical results. These explorations not only reinforce the statistical concepts being covered but also demonstrate how "what if" questions are analysed in the business world.

- **Problems:** Since experience working with real-world data is extremely important, each section ends with a set of problems. The problems cover a wide range of business topics and many use Canadian data. Students can use these problems to practise the procedures and templates given in the example.

- **Data disk:** A data disk provides the completed spreadsheets constructed in the examples. Students can use these files both to import the data required to work through the examples themselves and to check their results once they have completed the spreadsheets. The disk also contains data sets required for the problems.

ACKNOWLEDGMENTS

From Addison Wesley Longman, special thanks go to publisher Ron Doleman for his editorial insight and assistance, and to Suzanne Schaan for expertly editing the final version of the manuscript. Thanks to Thomas Hart for meticulously testing all the keystroke instructions. I also benefited from the professional attention of everyone involved from the production department. Any remaining mistakes are my own.

From McGill University's Faculty of Management, I am grateful to Brian Smith, Morty Yalovsky, J.L. Goffin, Derek Hart, and Dean Wallace Crowston, without each of whom I would have never been in the position to tackle such a project.

Finally, I thank my wife, Brenda, for her support in a very time-consuming process.

Introduction: Excel Basics

Microsoft Excel is the spreadsheet application for Microsoft Office. Excel is based on a concept called *workbooks*. A workbook is the electronic equivalent of a three-ring binder. Inside workbooks you will find sheets, such as worksheets and chart sheets. You can move or copy sheets between workbooks as well as rearrange sheets within a workbook. By entering information in the form of labels, values, and formulas into worksheet cells, you create *spreadsheets*, which are useful for summarizing, tabulating, and analysing data.

Each sheet's name appears in a set of page tabs running along the bottom of the worksheet window. By clicking the tabs, you can move from sheet to sheet. To change the name of a sheet, simply double-click on its page tab, then type the desired name. (The sheet names in the example files for this text have been named by the example numbers, and thus we will not add a step in every example to change the sheet name.)

At the top of the basic Excel worksheet window is the title bar, which shows the name of the program and the name of the file that is currently open. Just underneath, you'll find the menu bar, which provides pull-down menus that allow you to perform various worksheet tasks. Below the menu bar are the standard and formatting toolbars. The buttons on these toolbars perform tasks with one mouse click that otherwise would require several menu selections. In the line below the toolbars, the box on the left names the selected cell (the cell itself is also outlined in black). To the right, cell contents are edited in the formula bar.

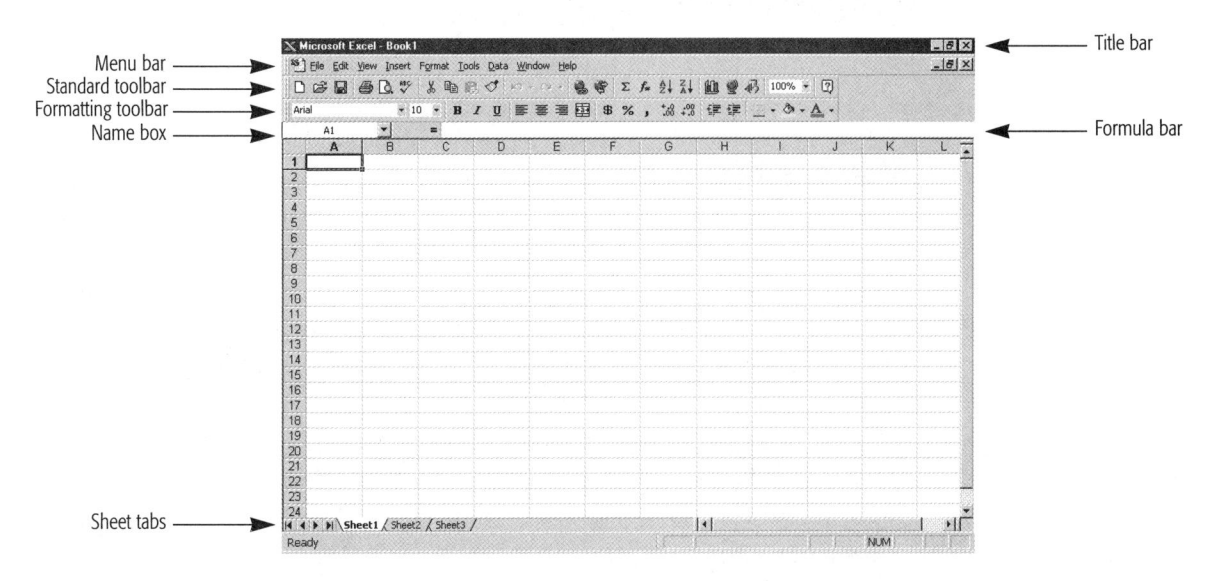

OPENING A FILE

To open a file, select **Open** from the **File** menu or click the **Open** icon on the standard toolbar. Select the directory within which the desired file is located. Click on the file name to select it, then click **Open**.

IMPORTING DATA INTO A WORKSHEET

Data can be copied and pasted from one Excel file into another Excel file. Simply open the file from which you want to copy and select the desired cells. Click the **Copy** button on the toolbar and then close the file. Open the file to which you would like to paste, select the top left cell of the new range, and click the **Paste** button on the toolbar. (This procedure should be used to import data when you work through the examples and problems in this book.)

Excel can also import data sets that have been stored in many different file formats used by other spreadsheet applications. One common format is text files, which contain unformatted data separated by delimiters (such as spaces or tabs) or aligned in fixed-width columns. The **Text Import Wizard** can be used to specify how the data in a text file are organized. To open the contents of a text file in a new Excel worksheet, open the desired file to launch the **Text Import Wizard**.

In Step 1 of the **Text Import Wizard**, select the **Delimited** or **Fixed Width** option button and click **Next**. In Steps 2 and 3, check that the data format appears correctly in the preview box. In most cases, you can accept the **Text Import Wizard**'s suggested formats for all columns by clicking the **Next** and **Finish** buttons, respectively. The file will then open.

At this point, you should check to see that the data have been correctly entered into all the columns. Select **Save As** from the **File** menu, and then click the **Save as type** drop-down arrow. Select **Microsoft Excel Workbook** from the list. Click **Save** to create an Excel file of the text.

SAVING YOUR WORK

It is advisable to save your work every two to three minutes. You don't want to lose hours of work for "unknown" reasons. The first time you save a new file, select **Save As** from the **File** menu. Choose a name for your file. Under the **Save as type** option, make sure that **Microsoft Excel Workbook** is selected. Finally, select the directory where you want to save your file. For future saves, simply click the **Save** button on the Standard toolbar or choose **Save** from the **File** menu. If you want Excel to automatically save your workbook at specified time intervals, choose the **AutoSave** command from the **Tools** menu. This command is an add-in, so if the command does not appear on your **Tools** menu, choose **Add-Ins** from the **Tools** menu, check **AutoSave**, then click **OK**.

COPYING AND PASTING

Copying and pasting can be done using the **Copy** and **Paste** buttons on the standard toolbar or by selecting **Copy** and **Paste** from the **Edit** menu. Select the cell or cells to be copied, then click **Copy**. A moving border appears around the cell(s) being copied. Move the cursor to the destination and click **Paste**.

- To paste a single cell to a block of cells: After copying, highlight the whole block to which you wish to paste, then click the **Paste** button.

- To paste a block of cells to another block of the same size: After copying, select the cell at the top left corner of the block to which you wish to paste, then click the **Paste** button.

- To paste a block of cells to a larger block of cells (for example, to copy one column to two other columns, or to copy a 2×3-cell block to a 4×6-cell block): After copying, highlight the whole block to which you wish to paste, then click **Paste**. (You will receive an error message if the block to which you try to paste is not a multiple of the block which you have copied.)

CELL FORMATTING

Excel's formatting options can be used to improve a spreadsheet's presentation, as different formats may be appropriate in different situations. (Note that the example files that accompany this text use Excel defaults except where changes have been specified in "Setting Up the Template.") The following are the most common formatting options:

- To change the width of a column to fit the labels or data, from the **Format** menu select **Column** then **Width**. Enter the desired width, then click **OK**.

- To make labels stand out from the data, block them by dragging the mouse, and then click the desired buttons on the formatting toolbar (for example, **Bold** or **Italic**).

- Excel's default is to align text on the left and numbers on the right. Alignment can be changed using the **Align Left, Center, Align Right,** and **Justify** buttons on the formatting toolbar.

- Numbers in a single cell or a range of cells containing numeric data can be formatted by highlighting the cells, selecting **Cells** from the **Format** menu, and then choosing the **Number** tab. The **Increase Decimal** and **Decrease**

Decimal buttons on the formatting toolbar can also be used to change the number of decimal places. Very small numbers and very large numbers are represented using scientific notation. For example, 0.000000000133967 appears as 1.33967E-10 and 9985746475857 appears as 9.98575E+12.

- You can also insert borders around cells or shade cells. First, block the desired range. From the **Format** menu, select **Cells**. Then select the **Border** or **Patterns** tab and format the cells as desired.

ENTERING FORMULAS

Inputting a *formula* into Excel is very simple. There are two basic things to remember:

1. A formula always begins with an equal sign (=).

2. A formula is not calculated until you hit the **Enter** key.

Formulas can combine constants with existing operators, references, functions, text, and names to produce new values or perform various other tasks. For example a formula may look like this:

$$=0.1118*(C7-B7)*1000-0.27*E7$$

Note that, after you hit **Enter,** the calculated value appears in the cell, but the formula still shows in the formula bar when that cell is selected, as seen in the display below.

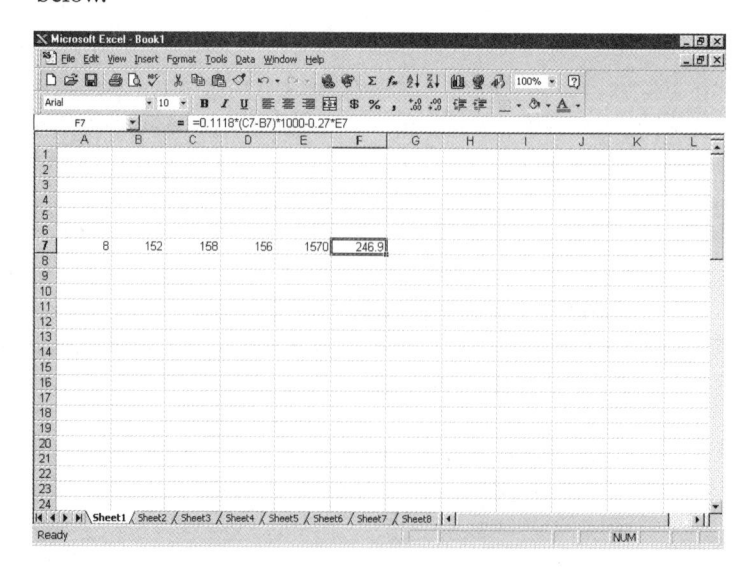

To edit a formula, place the cursor on the cell in which you typed the formula. Click on the formula in the formula bar and edit it as desired. When complete, click the check mark on the formula bar or hit **Enter**.

When calculating values in columns, after entering the first formula you simply have to copy the formula to the rest of the column. What allows you to do this is the fact that the cell addresses in your formula are *relative*, meaning that Excel adjusts them when the formula is moved. If you move the formula down one row, Excel moves all the formula cell addresses down one row as well. In the display above, the formula in Cell F7 references Cells C7, B7, and E7. In the display below, that formula has been copied down the column; note that Excel has moved the formula cell addresses down, so that the formula in Cell F8 references Cells C8, B8, and E8.

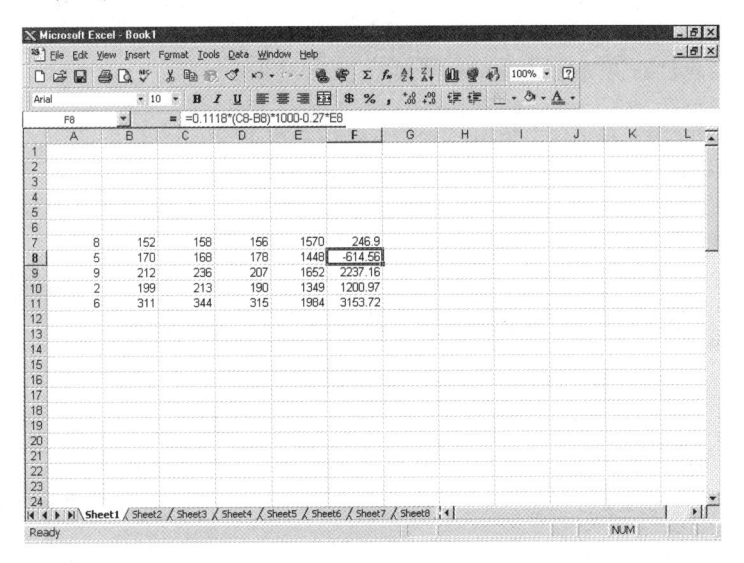

By typing a dollar sign in front of the column and row references (for example, **$A$12**), you "lock" them or make the address *absolute* rather than relative. Use this type of address when you don't want a cell address adjusted when a formula is copied (for example, if the cell contains the value of the mean of a data set). In *mixed* cell addresses, only one component (row or column) is absolute.

To reference a cell in a different sheet, the name of that sheet must be included in the cell reference, followed by an exclamation mark. For example, **Sheet1!A1** references the cell in the first column and row of the sheet named "Sheet1." This form allows you to distinguish between two similarly located cells of two different sheets (for example, **Sheet1!A1** and **Sheet2!A1**). When a cell address does not contain the name of a sheet, the current sheet is implied.

You may also need a function as part of your formula, whether you are summing a column of numbers or computing probabilities from the chi-square distribution. Each function has its own syntax, such as the following:

=AVERAGE(A2:A21) will return the mean of Cells A2 through A21.

=BINOMDIST(B6,8,0.25,0) will return the binomial distribution probability where B6 refers to the cell containing the number of successes, 8 is the number of trials, 0.25 is the probability of success on each trial, and 0 returns the discrete (rather than cumulative) probability.

Excel's **Paste Function** feature will help you add a function. Click the **Paste Function** button (f_x) on the standard toolbar, and then choose from the menu of function categories (one is **Statistical**). The **Paste Function** guides you through the function process so that you can easily finish its syntax.

DATA ANALYSIS TOOLS

Select **Data Analysis** from the **Tools** menu, and you will find most of the statistical tools that you will need for your statistics course. The dialog box that appears contains a scrollable list in which the first item is **Anova: Single Factor** and the last choice is **z-Test: Two Sample for Means**.

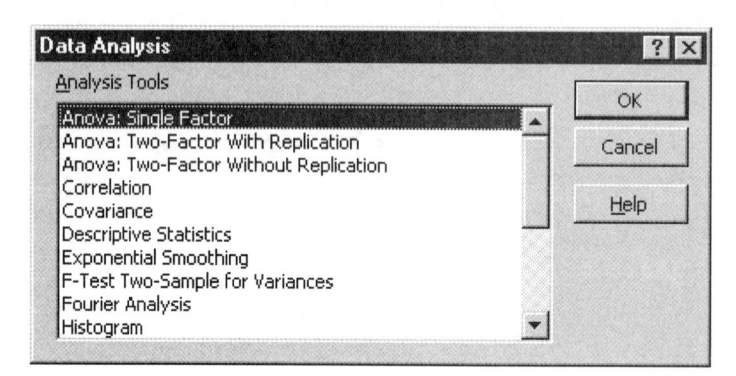

These are predefined routines that allow you to obtain results that are often very difficult or even impossible to obtain using simple formulas. If **Data Analysis** does not appear as a **Tools** menu choice, select **Add-Ins** from the **Tools** menu. Select the **Analysis ToolPak** check box and click the **OK** button. **Data Analysis** should now appear as a menu selection under the **Tools** menu. If **Analysis ToolPak** is not included in the list of available add-ins, it was not included when your copy of Excel was installed, and you must add it using the Excel setup program.

PREVIEWING AND PRINTING

Before printing, you can select **Page Setup** from the **File** menu to access numerous options that control the way your printed pages look. The **Page** tab contains **Orientation** buttons that allow you to choose whether your worksheets are printed vertically (**Portrait**) or horizontally (**Landscape**). You can reduce or enlarge the printed size of a worksheet or specify the number of pages to fit on one printed page with the **Scaling** options. The **Margins** tab allows you to set the size of margins on the printed page or to centre worksheets on a page. The **Header/Footer** tab allows you to change or customize both the header and the footer on a worksheet. Finally, the **Sheet** tab gives you the option of choosing whether gridlines and column/row headings are included in your printed pages. When you finish with **Page Setup**, you can preview your job by simply selecting **Print Preview** from the **File** menu.

If you wish to print only part of a spreadsheet, use the mouse to highlight the area to be printed. To print, select **Print** from the **File** menu. From here you can select the number of copies or what you want to print, be it a selection or a selected sheet of a whole workbook.

CHAPTER 1
Presenting Data in Basic Charts

1.1 CREATING PIE CHARTS

Quantitative data can be displayed in a *pie chart*. The entire circle is divided into slices that represent a number of categories. Each slice is proportional to the relative frequency of the category it represents. While bar and column charts explicitly show the frequency amounts, the pie chart shows the relative amounts as a percentage of the total. The pie chart must also include some information that identifies each slice of the pie. The steps for setting up a pie chart are shown in the following example.

EXAMPLE — **ABORIGINAL ORIGIN BY CITY** Chap01\Ex1-1.xls

The total population of aboriginal origin in 10 large Canadian cities (whole) is divided among the cities (portions):

City	Population
Winnipeg	45,705
Montreal	45,230
Vancouver	43,440
Edmonton	43,355
Toronto	40,555
Ottawa–Hull	31,220
Calgary	24,595
Saskatoon	14,530
Regina	13,055
Hamilton	11,245

Source: Statistics Canada; Indian and Native Affairs Canada.

Prepare a pie chart showing the population of aboriginal origin in each city as a proportion of the total population of aboriginal origin for the 10 Canadian cities.

Setting Up the Template

1. Type the given headings and data in Columns A and B or import the headings and data from file **Chap01\Ex1-1.xls**.

2. Select Cells A1:B1. From the **Format** menu, choose **Column, Width**. Change the width to **12**. To format the headings, select Cells A1:B1 and click **Bold** and **Center** on the toolbar.

3. Select Cells B2:B11. From the **Format** menu, choose **Cells, Number, Category: Number,** and click the **Use 1000 Separator (,)** check box.

4. Select Cells A2:B11 and then click the **Chart Wizard** button. Under **Standard Types**, select **Pie** for **Chart type**. Select the first **Chart sub-type** and click the **Next** button.

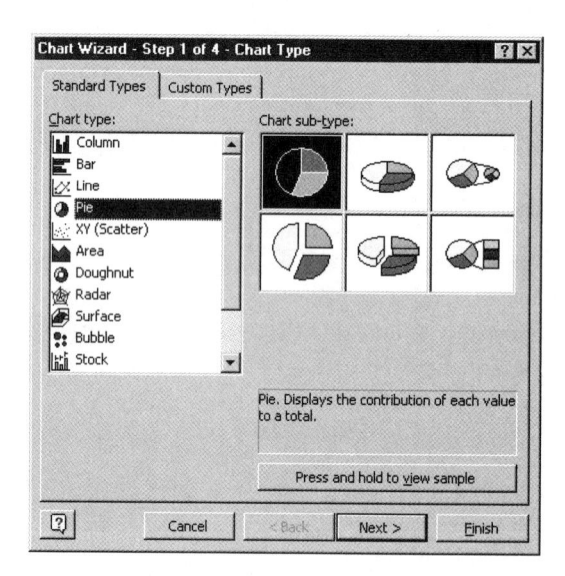

SECTION 1.1 • CREATING PIE CHARTS

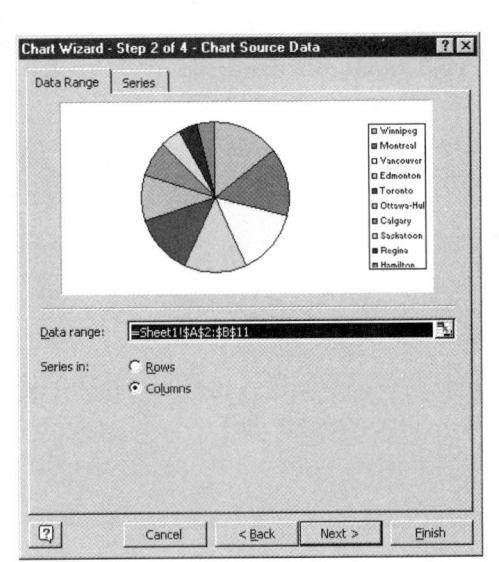

5. Make sure the data series is in **Columns** and click the **Next** button.

6. Under **Titles**, add **Chart Title: Aboriginal Origin by City**; under **Legend**, verify that the **Show legend** check box is selected. Finally, under **Data Labels**, select **Show percent**. Click the **Next** button.

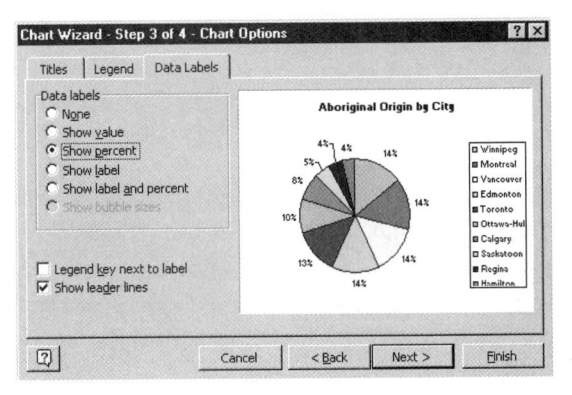

7. Place the chart as an object in the current sheet. Click the **Finish** button.

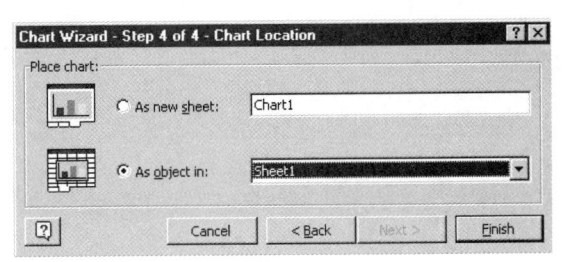

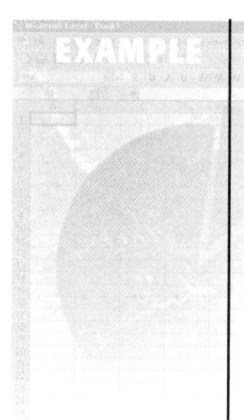

8. Click once anywhere within the chart box. When you place the mouse pointer on one of the eight square handles on the border, it changes to a double-arrow pointer. By clicking and dragging on any of these handles, enlarge the chart to improve its appearance. Release the mouse pointer when the desired size is achieved. To move the chart, click anywhere within the chart box and drag it to the desired location in the spreadsheet.

Editing/Formatting the Pie Chart

To edit a chart, click anywhere within the chart box. The frame of the chart should change and the menu at the top of the screen changes to those commands used by charts.

1. To move a label, click on it and drag it to the desired location. To edit a title, click to select it; click the text a second time to edit.

2. Click the pie once and all the slices will be selected. If you drag the mouse to the outside, then all the slices will explode. If you click twice on a particular slice, then that slice will be selected. You can explode the selected slice by dragging it to the outside.

3. To orient the slices as desired, double-click the pie and then select the **Options** tab. Change the angle of the first slice to the desired number of degrees from the vertical.

4. To change the colour of a slice, click on the desired slice and make certain that only that slice is selected and not the entire pie. Choose **Selected Data Point** under the **Format** menu and then select the **Patterns** tab. Customize the border and area as desired. Double-click the legend and chart area to edit them.

5. To select a new chart type, select **Chart Type** from the **Chart** menu. Choose the desired **Chart type** and **Chart sub-type** and click **OK**.

6. To format the percentages, double-click to select them. Select the **Number** tab and change the number of decimal places as desired. Click **OK**.

Tips and Notes

1. Pie charts will not be very informative if the number of slices is large and each of the slices does not represent a significant percent of the total (say, 5% or less).

2. If the data labels or percents are crowded and not readable, enlarge the chart as in Step 8 under "Setting Up the Template." A larger pie better reveals its smaller slices.

3. Remember that pie charts are updated as soon as data is changed.

4. If you notice a mistake in your chart when you are in one of the **Chart Wizard** dialog boxes, use the **Back** button to return to previous dialog boxes.

5. If you click outside the chart during editing, the chart will no longer be in edit mode. Click anywhere on the chart again so that you can modify it.

PROBLEMS

Chap01\P1-1#1.xls

1. Prior to a meeting on tax reform, a government researcher has gathered information on employees' average weekly earnings. Since there are several hundred different industries operating in Canada, the researcher selects a random sample of eight industries. The average weekly earnings for employees in those eight industries in November 1995 are recorded in the table below. Prepare a pie chart for these data.

Industry Group	$ (seasonally adjusted)
Forestry	751.65
Mining	1,002.14
Manufacturing	701.49
Construction	674.78
Transportation	728.00
Retail trade	349.62
Real estate	662.81
Business services	630.84

Source: Statistics Canada.

PROBLEMS

Chap01\P1-1#2.xls

2. Canadian university students faced significant increases in tuition for the 1995–96 academic year. The main reason for these increases was that universities have had to cope with a decline in both federal and provincial funding; therefore, the importance of student fees as a source of university income has been accentuated. In researching the trend, a student's society at Queen's University in Kingston selected a random sample of 10 Canadian universities and obtained the following data concerning tuition fees for full-time undergraduate arts students. Prepare a pie chart.

University	Undergraduate Arts Tuition ($)
Memorial University	2,312
University of Prince Edward Island	2,820
Dalhousie University	3,095
University of New Brunswick	2,610
University of Montreal	1,665
University of Toronto	2,451
Simon Fraser University	2,377
University of Western Ontario	2,550
University of Alberta	2,529
University of British Columbia	2,295

Source: Statistics Canada.

Chap01\P1-1#3.xls

3. Thousands of people across Canada are angry and frightened about what they see as an explosion of the homicide rate. Many Canadians blame the violence on American television networks seen in most Canadian homes, while others remain convinced that the criminal justice system is too soft on offenders. Homicide is often considered an urban phenomenon, since metropolitan areas with higher populations seem to report more homicides. The following table shows the number of homicides by metropolitan area for 1994. Prepare a pie chart.

Metropolitan Area	Homicides (1994)
Toronto	85
Montreal	75
Vancouver	48
Ottawa–Hull (Ontario)	12
Ottawa–Hull (Quebec)	5
Edmonton	24
Calgary	18
Quebec City	11
Winnipeg	18
Hamilton	13

Source: Statistics Canada.

1.2 CREATING BAR CHARTS AND COLUMN CHARTS

Quantitative data can be displayed in a *bar chart* or *column chart*. These charts explicitly show frequency amounts, and they are used interchangeably. The length of each bar represents the frequency (or percentage) of observations falling into a given category. In Excel, bar charts have horizontal bars and column charts have vertical bars. The steps for setting up a bar or column chart are shown in the following example.

EXAMPLE — TOP 10 TRADING PARTNERS

Chap01\Ex1-2.xls

Canada and the United States are the globe's biggest two-way traders. To put this fact in perspective, consider the following table depicting the world's top 10 bilateral trading partners in 1995:

Trading Partners	$U.S. billion
U.S.–Canada	285
U.S.–Japan	192
U.S–Mexico	126
Germany–France	114
Hong Kong–China	94
Germany–Italy	86
Germany–Netherlands	84
Germany–Britain	78
U.S.–Germany	68
Germany–Belgium	68

Source: Statistics Canada.

Prepare a bar chart with the trading relationship on the vertical axis and the value of goods traded ($U.S. billion) on the horizontal axis.

Setting Up The Template

1. Type the given headings and data in Columns A and B or import the headings and data from file **Chap01\Ex1-2.xls**.

2. Select Cells A1:B1. From the **Format** menu, choose **Column, Width**. Change the width to **20**. To format the headings, select Cells A1:B1 and click **Bold** and **Center** on the toolbar.

3. Select Cells A2:B11 and then click the **Chart Wizard** button. Under **Standard Types**, select **Bar** for **Chart type**. Select the first **Chart sub-type** and click the **Next** button.

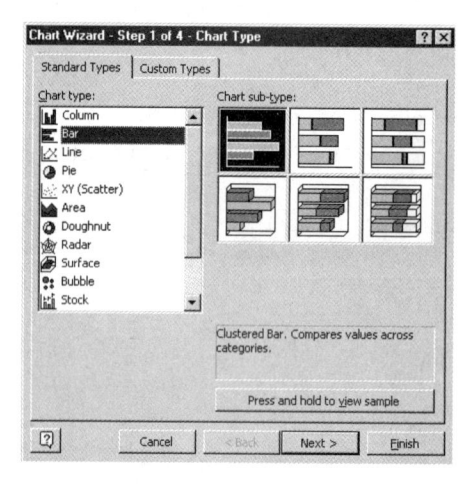

4. Make sure the data series is in **Columns** and click the **Next** button.

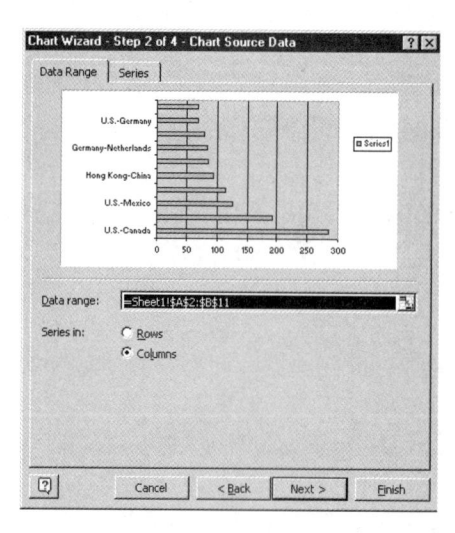

5. Under **Titles**, add titles for the chart, the X-axis, and the Y-axis by using the mouse or **Tab** key to navigate to the other fields. Under **Gridlines**, remove any major or minor gridlines by clicking away the appropriate check boxes. Under **Legend**, remove the legend by clicking away the check mark in the **Show legend** check box. Click the **Next** button.

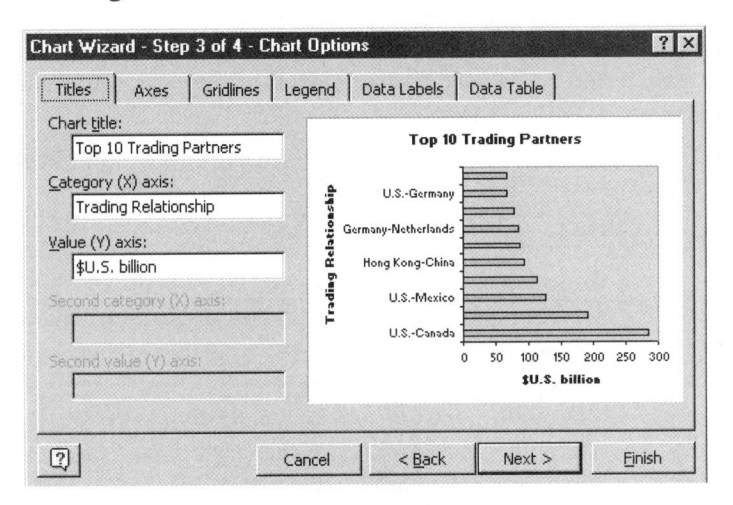

6. Place the chart as an object in the current sheet. Click the **Finish** button.

7. Click once anywhere within the chart box. When you place the mouse pointer on one of the eight square handles on the border, it changes to a double-arrow pointer. By clicking and dragging on any of these handles, enlarge the chart to improve its appearance and to display all the labels. Release the mouse pointer when the desired size is achieved. To move the chart, click anywhere within the chart box and drag it to the desired location in the spreadsheet.

Editing/Formatting the Bar Chart

To edit a chart, click anywhere within the chart box. The frame of the chart should change and the menu at the top of the screen changes to those commands used by charts.

1. To convert the bar chart to a column chart, select **Chart Type** from the **Chart** menu. Choose **Column** for **Chart type**, select the desired **Chart sub-type**, and click **OK**.

2. To move a label, click on it and drag it to the desired location. To edit a title, click to select it; click the text a second time to edit.

3. To change the width between the bars, double-click on the bars. Select the **Options** tab, change the value of the **Gap width**, and click **OK**. On the **Data Labels** tab, select **Show value** to place the frequency at the end of each bar.

4. The trading partners on the chart are in reverse order compared to the original table. Double-click on the vertical axis. On the **Scale** tab, select the check box for **Categories in reverse order** and click **OK**.

5. To change the colour of a bar, click on the desired bar and then click once more to make certain that only that bar is selected and not the entire set of bars. Choose **Selected Data Point** under the **Format** menu, and then select the **Patterns** tab. Customize the border and area as desired. To edit the plot area, double-click on it.

6. To format the horizontal axis, double-click on it. Select the **Number** tab and choose **Currency** in the category list box. Change **Decimal places** as required, and click **OK**.

Tips and Notes

1. All bars should be of the same width so as not to be misleading. Only the length of the bars should be different.

2. If the data labels are crowded and not readable, enlarge the chart as in Step 7 under "Setting Up the Template." A larger bar chart better reveals its labels.

3. Remember that bar charts are updated as soon as data is changed.

4. If you notice a mistake in your chart when you are in one of the **Chart Wizard** dialog boxes, use the **Back** button to return to previous dialog boxes.

5. If you click outside the chart during editing, the chart will no longer be in edit mode. Click anywhere on the chart again so that you can modify it.

PROBLEMS

Prepare bar and column charts for the scenarios in Section 1.1 (pp. 5–6).

1.3 CREATING LINE CHARTS AND FREQUENCY POLYGONS

LINE CHARTS

Sequence data or time series data can be displayed in a *line chart* so that patterns in the data can be identified. Although bar or column charts can be used for displaying a single time series, displaying multiple time series on the same bar or column chart is confusing; a line chart should be used in these situations. The steps for setting up a line chart are shown in the following example.

EXAMPLE: WATER TEMPERATURE BY HOUR

Chap01\Ex1-3.xls

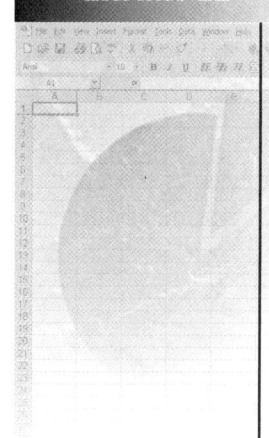

Based on concerns voiced by a local citizens' group, Environment Canada wishes to study the impact of a manufacturing plant built on the Peace River in northeastern Alberta. The citizens' group claims that the plant's activities are raising the temperature of the water in the river and thereby damaging the delicate aquatic ecosystems that have existed for thousands of years. In response, Environment Canada measured the water temperature hourly for four consecutive fall days at an area that is believed to be most affected. The following table gives sample temperature measurements in degrees Celsius, beginning at 12:00 a.m.:

Hour	Temperature
1	−11.54
2	−11.96
3	−13.37
4	−13.27
5	−13.62
6	−14.03
7	−14.36
8	−14.32
9	−14.45
10	−14.23
11	−14.50
12	−11.71
13	−10.29
14	−9.88
15	−8.74
⋮	⋮

(complete data on disk)

Prepare a line chart with the temperature on the vertical axis and the hour on the horizontal axis.

Setting Up the Template

1. Import the headings and data from file **Chap01\Ex1-3.xls** into Columns A and B.

2. Select Cells A1:B1. From the **Format** menu, choose **Column, Width**. Change the width to **10**. To format the headings, select Cells A1:B1 and click **Bold** and **Center** on the toolbar.

SECTION 1.3 • CREATING LINE CHARTS AND FREQUENCY POLYGONS   13

3. Select Cells A2:B97 and then click the **Chart Wizard** button. Under **Standard Types**, select **Line** for **Chart type**. Select the first **Chart sub-type** and click the **Next** button.

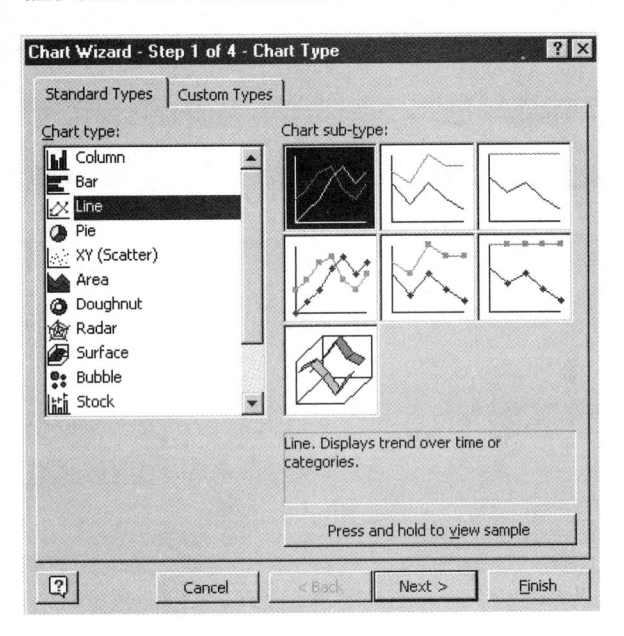

4. On the **Data Range** tab, make sure the data series is in **Columns**. On the **Series** tab, remove **Series1**. Click the **Next** button.

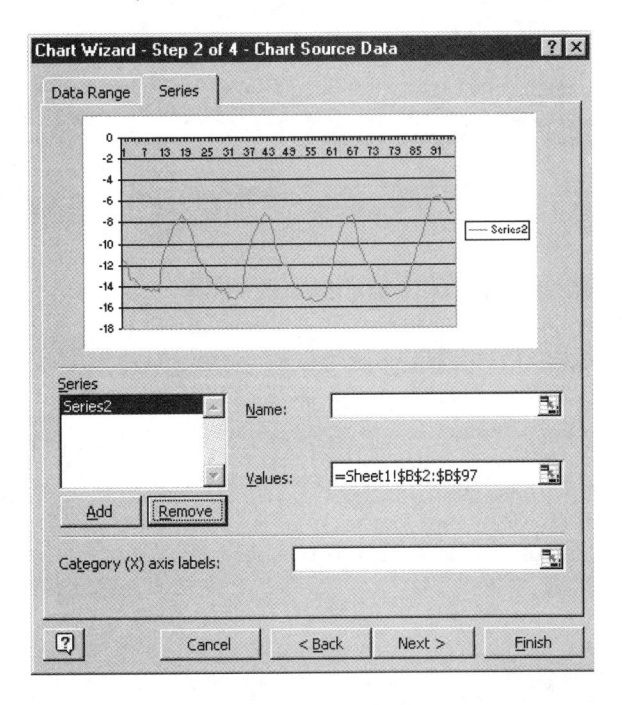

EXAMPLE

5. Under **Titles**, add titles for the chart and the Y-axis by using the mouse or **Tab** key to navigate to the other fields. Under **Gridlines**, remove any major or minor gridlines by clicking away the appropriate check boxes. Under **Legend**, remove the legend by clicking away the check mark in the **Show legend** check box. Click the **Next** button.

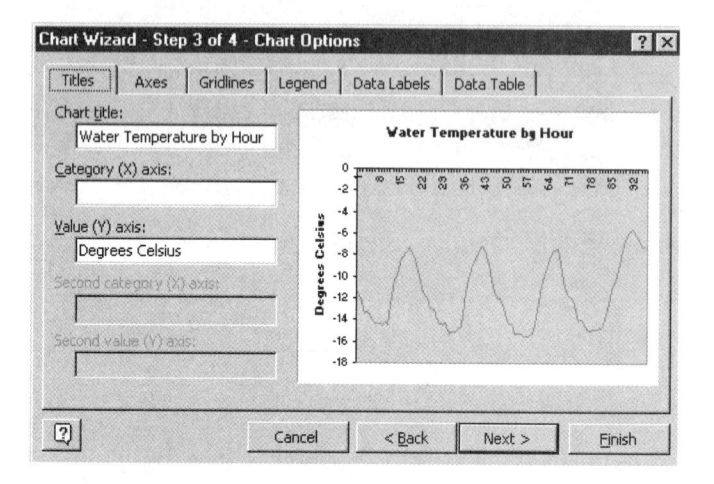

6. Place the chart as an object in the current sheet. Click the **Finish** button.

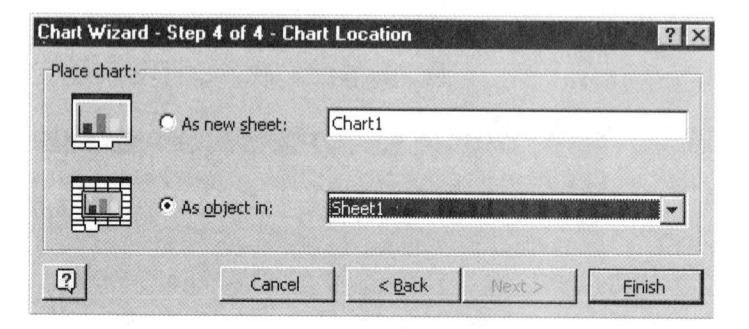

7. Click once anywhere within the chart box. When you place the mouse pointer on one of the eight square handles on the border, it changes to a double-arrow pointer. By clicking and dragging on any of these handles, enlarge the chart to improve its appearance. Release the mouse pointer when the desired size is achieved. To move the chart, click anywhere within the chart box and drag it to the desired location in the spreadsheet.

- For editing and formatting features, refer to Section 1.1 (pp. 4–5).

FREQUENCY POLYGONS

Line charts are also useful for displaying *frequency* or *relative frequency polygons* that represent the shape of a particular distribution. Polygons are formed with the midpoint of each class of the variable of interest on the horizontal axis and the frequency or percentage of observations per class on the vertical axis. The midpoints of each class are plotted at the height of the frequency for the interval and then they are joined with straight lines.

The frequency polygon is usually closed by adding an imaginary class interval with a frequency of zero at each end of the distribution and then extending a straight line to the midpoint of each of these classes. A graph of the relative frequencies would be the same as the graph of the absolute frequencies except that the values on the vertical axis would be in percent form. Otherwise, the construction of the line chart is exactly the same as described above with sequence data.

PROBLEMS

1. Data on the number of users (in 1,000s) of a Sears credit card during the period 1980 to 1996 follow:

Year	Users
1980	3.87
1981	3.99
1982	4.14
1983	4.28
1984	4.24
1985	4.25
1986	4.51
1987	4.60
1988	4.58
1989	4.68
1990	4.65
1991	4.72
1992	4.84
1993	4.81
1994	4.88
1995	4.93
1996	4.98

Prepare a line chart showing the year on the horizontal axis and users on the vertical axis. Does the time series contain a seasonal pattern? Explain.

PROBLEMS

Chap01\P1-3#2.xls

2. Consider the following data for quarterly sales of a cough medicine (in 1,000s) for Winnipeg over a seven-year period from 1987 to 1993:

	Quarter 1	Quarter 2	Quarter 3	Quarter 4
1987	12.6	15.2	18.8	16.0
1988	13.7	16.1	19.9	16.9
1989	14.6	17.2	20.8	18.0
1990	15.5	18.3	21.7	19.1
1991	16.4	19.4	22.6	20.2
1992	17.3	20.5	23.5	21.3
1993	18.2	21.6	24.4	22.2

Prepare line chart to represent any patterns in the data.

Chap01\P1-3#3.xls

3. A sample of dinner bills by table for 20 tables at the Vieux St. Charles restaurant in Quebec City has the following relative frequency distribution:

Dinner Bill ($)	Relative Frequency
39 to under 49	0.10
49 to under 59	0.30
59 to under 69	0.20
69 to under 79	0.20
79 to under 89	0.10
89 to under 99	0.10

Construct a relative frequency polygon for the distribution of dinner bills.

Chap01\P1-3#4.xls

4. The following is the distribution of annual income (in $1,000s) for a sample of 100 households in a Toronto suburb:

Annual Income	Number of Households
0 to under 10	5
10 to under 20	15
20 to under 30	20
30 to under 40	25
40 to under 50	15
50 to under 60	10
60 to under 70	5
70 to under 80	5

Construct a frequency polygon for the distribution of incomes.

PROBLEMS

Chap01\P1-3#5.xls

5. The relative frequency distribution for the number of overtime hours worked per month by a sample of 400 employees of the Montreal Urban Community Transit Corporation (MUCTC) is as follows:

Number of Overtime Hours	Relative Frequency
0 to under 4	0.05
4 to under 8	0.10
8 to under 12	0.20
12 to under 16	0.25
16 to under 20	0.28
20 to under 24	0.12

Construct a relative frequency polygon for the distribution of overtime hours.

1.4 CREATING SCATTER (*xy*) CHARTS

A *scatter (xy) chart* is a graph showing the shape and direction of the underlying relationship between two variables, x (independent or predictor variable) and y (dependent or response variable). Each point is plotted at its x–y coordinates. Also known as scatter plots or scatter diagrams, these charts are useful in examining the underlying relationship between two quantitative variables prior to a regression analysis.

EXAMPLE CANADA'S HOCKEY FORTUNES

Chap01\Ex1-4.xls

In the National Hockey League, does there appear to be a relationship between the number of Canadian teams that make the playoffs and the number of Canadian teams that finish with a winning percentage above .500 for the regular season? The following table shows the relevant information for the seasons from 1980–81 to 1995–96 inclusive:

EXAMPLE

Season	Canadian Teams over .500	Canadian Teams in Playoffs
1980–81	2	6
1981–82	3	6
1982–83	2	7
1983–84	3	6
1984–85	5	5
1985–86	4	7
1986–87	4	6
1987–88	3	5
1988–89	3	4
1989–90	4	5
1990–91	2	4
1991–92	3	4
1992–93	6	6
1993–94	4	4
1994–95	3	4
1995–96	1	5

Set up a scatter diagram showing the relationship between the number of Canadian teams over .500 and the number of Canadian teams in the playoffs.

Setting Up the Template

1. Type the given headings and data in Columns A, B, and C or import the headings and data from file **Chap01\Ex1-4.xls**.

2. Select Cells A1:C1. From the **Format** menu, choose **Column, Width**. Change the width to **10**. To format the headings, select Cells A1:C3 and click **Bold** and **Center** on the toolbar.

3. Select Cells B4:C19 (do not include the titles) and then click the **Chart Wizard** button on the toolbar. Under **Standard Types**, select **XY (Scatter)** for **Chart type**. Select the first **Chart sub-type** and click the **Next** button.

SECTION 1.4 • CREATING SCATTER (xy) CHARTS ◀ 19

EXAMPLE

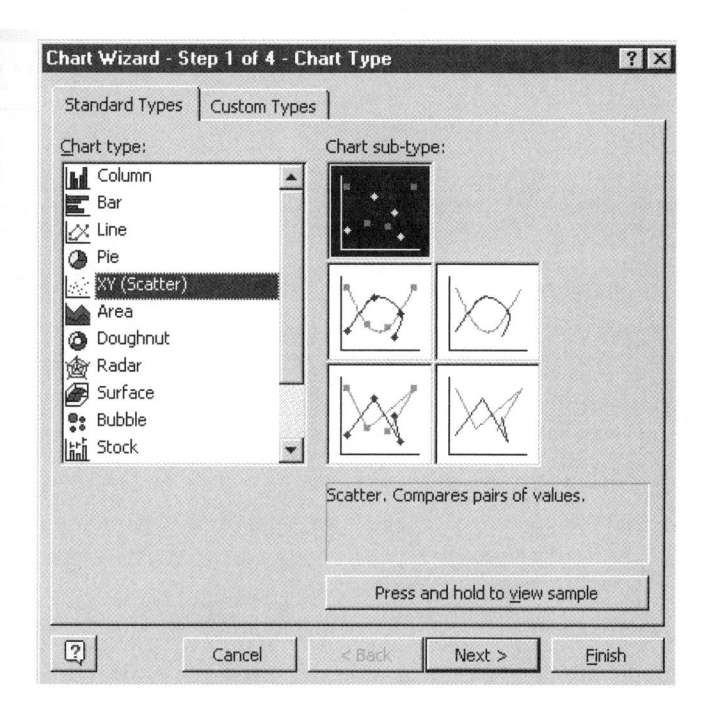

4. Make sure the data series is in **Columns** and click the **Next** button.

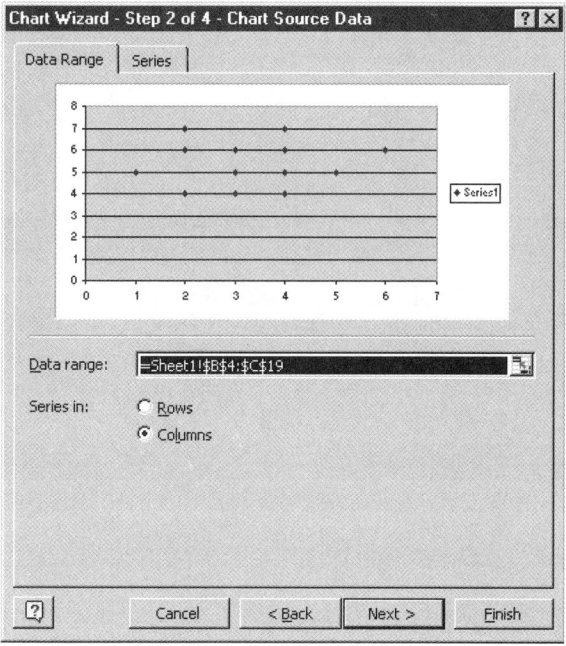

5. Under **Titles**, add titles for the chart, the X-axis, and the Y-axis by using the mouse or **Tab** key to navigate to the other fields. Under **Gridlines**, remove any major or minor gridlines by clicking away the appropriate check boxes. Under **Legend**, remove the legend by clicking away the check mark in the **Show legend** check box. Click the **Next** button.

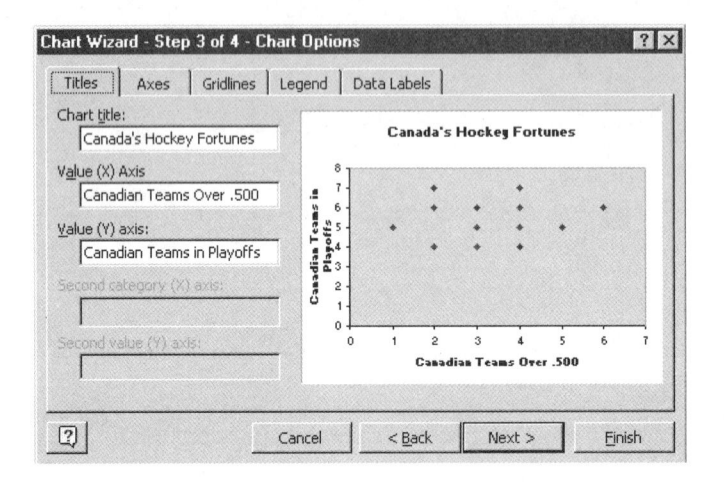

6. Place the chart as an object in the current sheet. Click the **Finish** button.

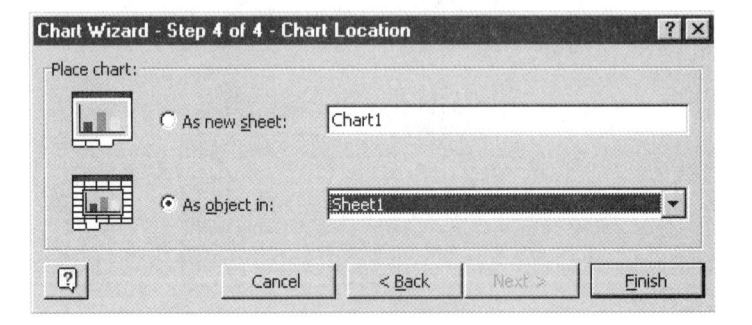

7. Click once anywhere within the chart box. When you place the mouse pointer on one of the eight square handles on the border, it changes to a double-arrow pointer. By clicking and dragging on any of these handles, enlarge the chart to improve its appearance. Release the mouse pointer when the desired size is achieved. To move the chart, click anywhere within the chart and drag it to the desired location in the spreadsheet.

We can edit the chart in much the same way as we edited the pie chart in Section 1.1 (pp. 4–5). To change the scale of an axis, double-click on that axis, choose the **Scale** tab, and change the **Minimum**, **Maximum**, **Major**, and **Minor** units as desired. Double-click on the plot area, titles, or data series to edit or format further. To move a label, click on it and drag it to the desired location.

Discussing the Outcome

We might expect that the greater the number of Canadian teams with winning records (over .500 winning percentage) in a given season, the greater the number of Canadian teams making the playoffs in that year. An examination of the scatter (*xy*) chart indicates clearly that there does not appear to be a positive or direct linear relationship (correlation) between the two variables, and therefore a regression analysis may not be useful. The points are scattered all over the graph and seem to fall along a straight line. If there were a positive relationship between *x* and *y*, then the two variables would increase and decrease together. If there were a negative (inverse) relationship between *x* and *y*, then the two variables would move in opposite directions.

The reason there does *not* appear to be a linear relationship between the two variables is that factors other than the number of Canadian teams with winning records influence the number of Canadian teams that make the playoffs in a given season. For example, the strength of the more numerous American teams in the league is an influencing factor. In the 1982–83 season, only two Canadian teams had winning seasons, but seven made the playoffs. This difference could be due in part to underachieving American teams.

PROBLEMS

Chap01\P1-4#1.xls

1. The manager of an assembly line is studying the relationship between the number of weeks of experience in wiring an electronic component and the number of faultless components produced during the past week. The manager conducted an experiment by selecting 10 workers at random. The following results were recorded:

Experience in Weeks (x)	Number of Components (y)
37	122
32	107
29	99
35	110
33	113
31	104
35	116
34	114
32	106
37	120

Prepare a scatter (*xy*) chart and comment on the direction and strength of the linear relationship between the two variables.

2. In the aftermath of the great ice storm of 1998 in southern Quebec and Ontario, an insurance company studied the relationship between the insured person's income (in $10,000s) and the amount of home insurance carried by the person (in $10,000s). The following data was randomly selected from the company's files:

Annual Income (x)	Amount of Insurance (y)
3.2	2.7
2.0	2.4
4.6	3.9
2.6	2.1
7.2	5.3
2.4	2.5
5.5	4.3
2.1	2.0
4.4	3.4
4.7	3.8
5.9	4.6
6.3	5.3

Prepare a scatter (*xy*) chart. Does there appear to be a positive or negative rela-

PROBLEMS

Chap01\P1-4#3.xls

tionship between the two variables?

3. Consumer awareness of a new product on the market can be measured by the percent of the target market that has heard about the product 30 weeks after it has been introduced. A market analyst interested in the relationship between advertising expenditure (in $1,000s) and consumer awareness (%) selected a random sample of 10 new products that have been on the market for 30 weeks. The following results were observed:

Advertising Expenditure (x)	Consumer Awareness (y)
220	47
160	24
110	13
820	88
630	67
470	62
690	70
180	31
170	44
340	53

Prepare a scatter (xy) chart and comment on the direction and strength of the linear relationship between the two variables.

CHAPTER 2
Describing Numerical Data

2.1 SORTING DATA INTO ORDERED ARRAYS AND FREQUENCY DISTRIBUTIONS

ORDERED ARRAYS

In Excel, the **Data, Sort** command will sort a data set into an *ordered array*, as illustrated in the following example.

EXAMPLE **LONG-DISTANCE PHONE BILLS** Chap02\Ex2-1.xls

The following represents a sample of 200 Sprint Canada long-distance phone bills (in dollars):

Bills
96.34
94.24
107.00
118.58
66.16
15.37
7.46
67.81
0.00
22.64
3.81
72.66
7.24
⋮

(complete data on disk)

Sort the data into ascending order.

Setting Up the Template

1. Import the heading and data from Column A of the sheet titled **Data** in file **Chap02\Ex2-1.xls**.

2. To format the heading, select Cell A1 and click **Bold** and **Center** on the toolbar.

3. Click on any cell in Column A and then, from the **Data** menu, choose **Sort**.

4. The **Sort** dialog box appears with the column heading at the top of the dialog box. The data can be sorted by one, two, or three variables; for each variable, the sorting can follow ascending or descending order. Select the **Header row** button to sort the data without moving the column heading(s). Click **OK**. The data will be sorted as shown on the sheet titled **Sort** in the file **Chap02\Ex2-1.xls**.

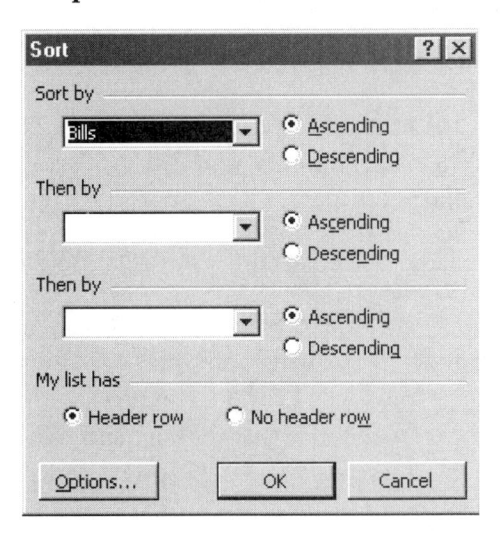

FREQUENCY DISTRIBUTIONS

A *frequency distribution* is a table showing the number of occurrences of a variable over the range of possible values, which are either considered individually or grouped into a set of mutually exclusive and exhaustive classes. Therefore, each and every value in a data set can be placed in one, and only one, of the class intervals. This method will summarize and describe very large data sets without loss or distortion of its characteristics. As a general rule, all the class widths should be the same and the width used should be convenient to work with. The steps for creating a frequency distribution using the **FREQUENCY** function are shown in the following example.

EXAMPLE

LONG-DISTANCE PHONE BILLS

Chap02\Ex2-1.xls

Once again, consider the sample of 200 Sprint Canada long-distance phone bills. Construct a frequency distribution with the following upper class boundaries:

| 14.99 | 29.99 | 44.99 | 59.99 | 74.99 | 89.99 | 104.99 | 119.99 |

The value 14.99 is used as an approximation of less than 15, 29.99 as an approximation of less than 30, and so on.

Setting Up the Template

1. Type or import the heading and data from Column B of the worksheet **Sort** in file **Chap02\Ex2-1.xls** into Column B of the worksheet containing the sorted phone-bill data.

2. To format the heading, select Cell B1 and click **Bold** on the toolbar.

3. Select Cells C2:C9 and type the function **=FREQUENCY(A2:A201,B2:B9)**. While holding down the **Control** and **Shift** keys, hit the **Enter** key. The frequencies now appear in Cells C2:C9. If you review the formulas just entered in Cells C2:C9, you will notice that they are enclosed in curly brackets { }, which indicate that these cells contain a special type of formula that cannot be individually edited.

4. Enter an appropriate title for Column C and format as desired.

Discussing the Outcome

The frequency distribution indicates that there are 69 bills less than $15, 35 bills between $15 and $30, and so on. Notice that Excel allows us to observe the effect of changes in the data on the frequency distribution. For example, we can change the third bill in the unsorted data set from $107.00 to $7.00. We now observe that the frequency for the first class has increased from 69 to 70 and the frequency of the last class has decreased from 17 to 16.

PROBLEMS

1. A sales manager has recorded the dollar values, to the nearest $10, of orders obtained by 32 salespeople in the last week. The results are as follows:

450	380	280	330	390	310	340	240
260	370	460	480	320	320	330	410
330	290	280	510	250	230	420	340
380	430	370	360	360	310	390	350

Sort the data set if desired. Construct a frequency distribution with six classes, each of width 50. The first class should be "220 to under 270."

2. A controller has recorded the following observations on the random variable "profit per retail sale (in dollars)":

68	38	54	88	62	58	76	34
52	60	70	42	72	48	62	54
50	82	30	52	60	56	42	64
78	50	44	36	64	58	48	52

Sort the data set if desired. Construct a frequency distribution with six classes, each of width 10. The first class should be "30 to under 40."

3. The following data show the list prices of a sample of houses for sale advertised in the Montreal *Gazette*. Prices are expressed to the nearest $1,000 (that is, the first two were listed for $108,900 and $129,000 respectively). The cities of Beaconsfield and Brossard were sampled with the following results:

Beaconsfield
108.9 129 139 149 149.9 184.5 189 239 274.9 340 369 525

Construct a frequency distribution with five classes, each of width 85. The first class should be "105 to under 190."

Brossard
79.9 92.5 104 125 128 129 129 135 135 165 229 299 325 329

Construct a frequency distribution with five classes, each of width 50. The first class should be "79.5 to under 129.5."

2.2 USING EXCEL FUNCTIONS TO PRODUCE DESCRIPTIVE STATISTICS

Descriptive statistics deal with summarizing, organizing, and presenting statistical data. The goal is to show important characteristics of the data without drawing any conclusions. Excel functions compute measures of central tendency, dispersion, and skewness for quantitative variables. The steps for computing descriptive statistics using Excel functions are shown in the following example.

EXAMPLE — HOSPITAL BED COUNTS

Chap02\Ex2-2.xls

The following table presents the percentage change in hospital bed counts for the 10 Canadian provinces between 1986 and 1995:

Province	Percentage Change in Hospital Bed Counts 1986–95
Newfoundland	−19.1
Prince Edward Island	−32.1
Nova Scotia	−34.8
New Brunswick	−34.1
Quebec	−29.0
Ontario	−27.1
Manitoba	−13.2
Saskatchewan	−35.7
Alberta	−53.5
British Columbia	−20.2

Source: Statistics Canada.

Use Excel functions to find the following sample statistics: sum, count, mean, median, mode, maximum, minimum, range, variance, standard deviation, skewness, kurtosis, 25th percentile, 75th percentile, interquartile range, and coefficient of variation.

Setting Up the Template

1. Type the given headings and data in Columns A and B or import the headings and data from file **Chap02\Ex2-2.xls**.

2. Select Cells A1:D1. From the **Format** menu, choose **Column, Width**. Change the width to **20**. To format the headings, select Cells A1:B5 and click **Bold** and **Center** on the toolbar.

3. Type the heading **Sample Statistics** in Cell C1, and type labels for the statistics listed in the example above into Cells C3:C18, or import the labels from Column C of file **Chap02\Ex2-2.xls**. Format as desired.

4. In Cell D3, enter the function **=SUM(B6:B15)** to compute the total of the data set.

5. In Cell D4, enter the function **=COUNT(B6:B15)** to compute the sample size.

6. In Cell D5, enter the function **=AVERAGE(B6:B15)** to compute the sample mean of the data set.

7. In Cell D6, enter the function **=MEDIAN(B6:B15)** to compute the sample median of the data set.

8. In Cell D7, enter the function **=MODE(B6:B15)** to compute the sample mode of the data set.

9. In Cell D8, enter the function **=MAX(B6:B15)** to display the maximum observation of the data set.

10. In Cell D9, enter the function **=MIN(B6:B15)** to display the minimum observation of the data set.

11. In Cell D10, enter the function **=D8-D9** to compute the range of the data set.

12. In Cell D11, enter the function **=VAR(B6:B15)** to compute the sample variance of the data set.

13. In Cell D12, enter the function **=STDEV(B6:B15)** to compute the sample standard deviation of the data set.

14. In Cell D13, enter the function **=SKEW(B6:B15)** to compute the skewness measure of the data set.

15. In Cell D14, enter the function **=KURT(B6:B15)** to compute the kurtosis measure of the data set.

16. In Cell D15, enter the function **=PERCENTILE(B6:B15,0.25)** to compute the 25th percentile of the data set.

17. In Cell D16, enter the function **=PERCENTILE(B6:B15,0.75)** to compute the 75th percentile of the data set.

18. In Cell D17, enter the formula **=D16-D15** to compute the interquartile range of the data set.

19. In Cell D18, enter the formula **=D12/D5*100** to compute the coefficient of variation of the data set.

20. To round the sample statistics, select Cells D5:D18 and click the **Decrease Decimal** button on the toolbar to change the number of decimals to 2.

2.3 USING THE DATA ANALYSIS TOOL TO PRODUCE DESCRIPTIVE STATISTICS

The **Data Analysis: Descriptive Statistics** tool computes measures of central tendency, dispersion, and skewness for quantitative variables. The steps for computing descriptive statistics using the **Data Analysis** tool are shown in the following example.

EXAMPLE HOSPITAL BED COUNTS

Chap02\Ex2-3.xls

Refer to the example in Section 2.2, which presented the percentage change in hospital bed counts for the 10 Canadian provinces between 1986 and 1995:

Province	Percentage Change in Hospital Bed Counts 1986–95
Newfoundland	−19.1
Prince Edward Island	−32.1
Nova Scotia	−34.8
New Brunswick	−34.1
Quebec	−29.0
Ontario	−27.1
Manitoba	−13.2
Saskatchewan	−35.7
Alberta	−53.5
British Columbia	−20.2

Source: Statistics Canada.

Use the **Data Analysis** tool to find the descriptive statistics.

Setting Up the Template

1. Type the given headings and data in Columns A and B or import the headings and data from file **Chap02\Ex2-3.xls**.

2. Select Cells A1:B1. From the **Format** menu, choose **Column, Width**. Change the width to **20**. To format the headings, select Cells A1:B5 and click **Bold** and **Center** on the toolbar.

3. From the **Tools** menu, choose **Data Analysis**. A dialog box containing a scrollable list of data analysis tools will appear. If **Data Analysis** does not appear, choose **Add-Ins** from the **Tools** menu. Check **Analysis ToolPak** in the scrollable list of available add-ins and click **OK**. **Data Analysis** should now appear as a menu selection under **Tools**. (If **Analysis ToolPak** is not included in the list of available add-ins, then you must add it using the Excel Setup program, as it was not included when your copy of Excel was installed.)

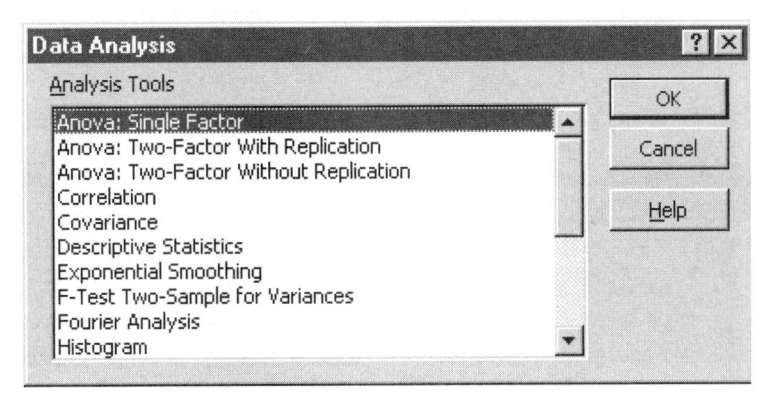

4. From the **Data Analysis** tools list, choose **Descriptive Statistics** and click **OK**. Enter **B6:B15** for the **Input Range** to indicate the location of the data. Verify that the **Grouped By: Columns** button is selected. **Labels in First Row** does not need to be checked, as we did not include the titles in our input range. Select the **Output Range** button then click in the box and enter **D1** for the range. Check **Summary Statistics** to produce a complete output of descriptive statistics. Check **Kth Largest** and **Kth Smallest** and enter **1** for K to display the maximum and minimum observations in the data set. Then click **OK** to have Excel produce the descriptive statistics.

32 ▸ CHAPTER 2 • DESCRIBING NUMERICAL DATA

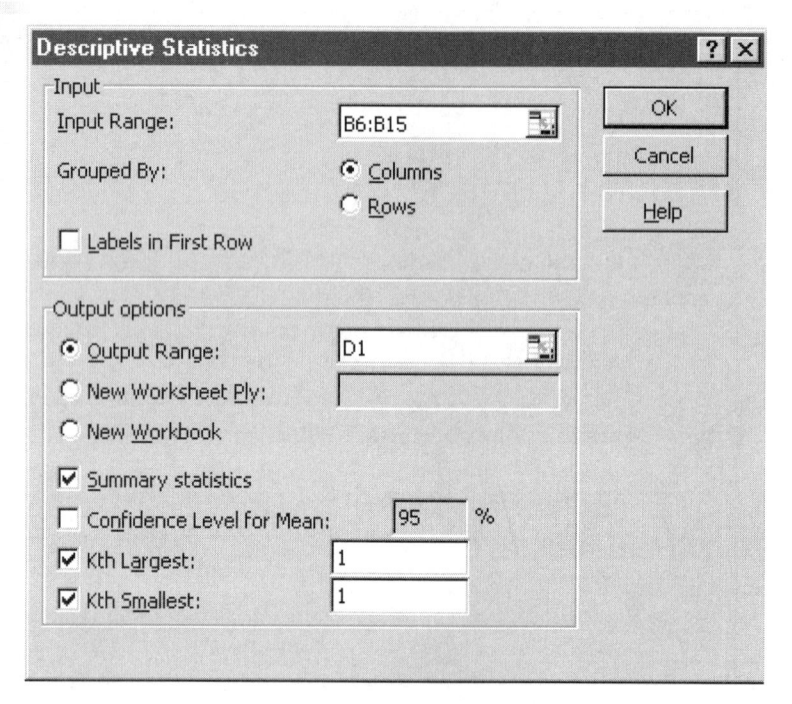

[Descriptive Statistics dialog box:
- Input Range: B6:B15
- Grouped By: ● Columns, ○ Rows
- ☐ Labels in First Row
- Output options:
  - ● Output Range: D1
  - ○ New Worksheet Ply:
  - ○ New Workbook
- ☑ Summary statistics
- ☐ Confidence Level for Mean: 95 %
- ☑ Kth Largest: 1
- ☑ Kth Smallest: 1
- Buttons: OK, Cancel, Help]

5. To improve the appearance of the output, select the whole output range and choose **Column** then **AutoFit Selection** from the **Format** menu. Click **OK**. Use the **Increase Decimal** and **Decrease Decimal** buttons on the toolbar as desired.

Discussing the Outcome

The advantage of using the **Data Analysis** tool rather than the various Excel functions to produce descriptive statistics is that many measures of central tendency, dispersion, and skewness can be produced with only one command. The output is now part of the spreadsheet and can be edited and formatted just like any other area of a spreadsheet.

The disadvantage of the **Data Analysis** tool is that, if the data set changes after the tool has been used, the output is not updated to reflect the changes. Scroll down the results in the output and notice that they exist only as numbers and not as functions in the spreadsheet. Steps 3 and 4 in "Setting Up the Template" would have to be repeated with the new data set in order to produce descriptive statistics reflecting the changes in the data. Producing descriptive statistics with Excel functions allows the results to be automatically updated as changes are made in the data set, but more commands are required to produce the results.

2.4 INTERPRETING DESCRIPTIVE STATISTICS

The sample *mean* is the sum of the members of a set of observations divided by the number of observations in the set. In the example in Sections 2.2 and 2.3, the mean percentage change is −29.88, calculated by dividing the *sum* of −298.8 by the *count* of 10.

The sample *median* is the value of the observation exactly at the centre of an ordered set of observations. It is therefore the value above which 50% of the data lie and below which the other 50% of the data lie. It is first necessary to find the median observation:

$$\text{Median position} = (n + 1)/2$$

The median is the middle value of the *ordered* list if n is odd and it is the average of the two middle values of the ordered list if n is even. In the example, the median is −30.55, which is the average of the fifth (−32.1) and sixth (−29) values of the ordered list, since $(n + 1)/2 = 5.5$.

The sample *mode* is the value in the data set that occurs most often. In the example, the mode function displays **#N/A** because there is no mode; all of the values in the data set occur just once. When a data set has multiple modes, Excel would display the value that first appears in the data set.

Skewness refers to the shape of the distribution of values. Positive (right) skewness indicates a distribution with an asymmetrical tail extending to more positive values and negative (left) skewness indicates a distribution with an asymmetrical tail extending to more negative values. In the example, the skewness measure is negative (−0.6624), because Alberta's percentage change of −53.5 is a larger drop compared to the other provinces. The median is not affected by such extreme observations and therefore is a preferred measure of central tendency for a data set that is skewed.

Kurtosis refers to the relative peakedness or flatness of a distribution compared to the normal (bell-shaped) distribution. Positive kurtosis indicates a relatively peaked distribution and negative kurtosis indicates a relatively flat distribution. In the example, the measure is positive (1.326).

The *range* of a data set is defined as the difference between the largest and smallest observations in the data set. In the example, the range of 40.3 is the *maximum* of −13.2 minus the *minimum* of −53.5. The problem with the range is that it is affected by extreme observations and does not use all the information in the data set it describes.

The sample *variance* is the average of the squared deviations of the observations from the mean. If the observations are tightly crowded around the mean, then the variance will be relatively small. If the observations are widely spread out around the mean, then the variance will be relatively large.

$$s^2 = \frac{\Sigma (x - \bar{x})^2}{n - 1}$$

In the example, the variance is 126.15. The variance is very sensitive to extreme observations. The calculation of the variance involves squaring the original units of the variable of interest. It is difficult to interpret the measurement in this form; standard deviation is more useful.

The *standard deviation* is computed by taking the positive square root of the variance. This operation converts the measure back into the same units as the original observations, just as the mean is.

$$s = +\sqrt{s^2}$$

In the example, the standard deviation is 11.23. If a constant were added or subtracted from every observation in a data set, the dispersion of the observations would not be influenced, and therefore the variance and standard deviation would remain unchanged. To compute the population variance and population standard deviation with n instead of $n - 1$ in the denominator, use the **VARP** and **STDEVP** functions.

The variance of data sets can only be compared when the same units of measurement are employed in the data sets and when the means of the data sets are approximately equal. When either of these conditions is not met, the *coefficient of variation* must be used to compare the variability in data sets, as it standardizes the unit of measure. The coefficient of variation is the standard deviation as a percentage of the mean and therefore is independent of the unit of measurement.

$$CV = 100(s/\bar{x})$$

In the example, the coefficient of variation is computed to be $100(11.23/-29.88) = -37.59$.

The *standard error* of the mean is equal to the sample standard deviation divided by the square root of the sample size. In the example in Section 2.3, it is 3.55. It will be used for inferential statistics discussed in Chapters 6 and 7.

The "Largest(1)" and "Smallest(1)" values in the example in Section 2.3 once again display the maximum and the minimum values in the data set. Had the

"Largest(3)" and "Smallest(3)" values been specified, Excel would have displayed the third largest and third smallest values in the data set.

For a given set of observations, the *Pth percentile* is the value below which P% of the observations lie. The *interquartile range* describes the range of the middle 50% of the observations in a data set. It is therefore the difference between the first and third quartiles (the 75th and the 25th percentiles).

$$\text{Interquartile range} = P_{75} - P_{25}$$

In the example, the 25th percentile is −34.625, which implies that approximately 25% of the data lie below −34.625. The 75th percentile is −21.925, which implies that approximately 75% of the data lie below −21.925. The interquartile range of 12.7 implies that the *middle* 50% of the observations in the data set span 12.7%.

PROBLEMS

1. The following data present the cash flow figures for a sample of companies in the automobile and telecommunications industries:

Automobile ($1,000,000s)	Telecommunications ($1,000,000s)
269.95	17.56
8.16	25.31
18.97	19.29
153.75	44.57
4.50	11.61
33.68	18.12
91.72	14.05
99.40	23.56
2.34	13.13
26.79	86.70

   a. Find the sample mean and sample standard deviation for both industries.
   b. Are the two data sets fairly symmetrical? Explain.
   c. Determine the range for each of the two industries.
   d. Compare the level of volatility in cash flows between the two industries. Which industry is the most volatile?

2. Parking meters in Edmonton were tested. A random sample of 10 meters on the west side of streets and 12 meters on the east side of streets showed the following periods of operation (in minutes) before indicating a parking violation when

set to allow 30 minutes of parking time:

| West side | 29 | 31 | 24 | 29 | 33 | 29 | 27 | 28 | 32 | 38 | | |
| East side | 35 | 26 | 38 | 23 | 38 | 26 | 37 | 25 | 25 | 26 | 36 | 37 |

a. Calculate the mean, median, and mode for each set of sample data.
b. If the fine for a parking violation is $100, on which side of the street would you prefer to park?
c. For the east side, what percentage of data falls within one standard deviation of the mean?
d. For the west side, what percentage of data falls within two standard deviations of the mean?
e. Calculate the interquartile range and coefficient of variation for each set of sample data.

3. An employee may choose to go to work by public transportation or by automobile. Sample times for each method are shown (in minutes):

| Public transportation | 25 | 28 | 31 | 38 | 32 | 27 | 28 | 30 | 42 | 35 |
| Automobile | 28 | 31 | 32 | 33 | 37 | 36 | 30 | 32 | 31 | 34 |

a. Compute the sample mean time to get to work for each method.
b. Compute the sample standard deviation for each method.
c. Are the data sets fairly symmetrical? Explain.
d. Based on your results above, which method of transportation should be preferred? Explain.
e. Compute the 80th percentile for each method.

4. A group of 50 children includes 7 one-year-olds, 18 two-year-olds, and 16 three-year-olds; all the rest are four-year-olds. The median age of this group of children is _____ and the distribution is skewed to the _____.

5. A sample of British Columbia's 1993 and 1994 exports of 50 commodities are recorded in the table below. Using summary measures, present a description of the information provided by the data in each year. Also, present a comparison of the data between the two years. Has the central tendency or the variation of the data changed from 1993 to 1994? Do any observations lie outside three standard deviations away from the mean?

Commodities	1993 ($1,000s)	1994 ($1,000s)
Logs, poles, and pulpwood	191,454	130,361
Lumber (softwood)	6,560,104	7,636,816
Cedar shakes and shingles	243,191	211,417
Softwood siding and moulding	178,626	184,730
Plywood	95,772	98,735
Pulpwood chips	100,222	80,101
Pulp	2,352,283	3,524,590
Newsprint	1,168,116	1,133,390
Paper and paperboard	482,965	600,447
Converted paper	80,769	108,223
⋮	⋮	⋮

(complete data on disk)

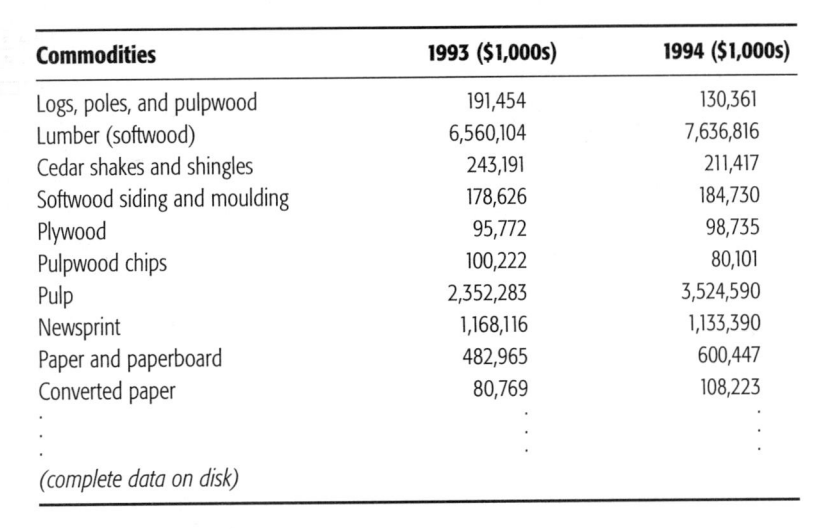

Source: Statistics Canada.

2.5 CREATING HISTOGRAMS

To create a *histogram*, classes are plotted on the horizontal axis and frequencies are plotted on the vertical axis. Adjacent bars of equal width are drawn to the height that corresponds to the frequency of that interval. The horizontal boundaries of each vertical bar correspond to the class limits. Histograms are useful in giving a general idea of the shape of a distribution.

The **Data Analysis: Histogram** tool can be used to generate a histogram for a data set, as well as a frequency distribution and cumulative relative frequency polygon. A cumulative relative frequency polygon is a graph of the cumulative relative frequencies on the vertical axis and the upper class limits on the horizontal axis. The class limits should be determined before using the data analysis tool, to make them easy to analyze. If not, Excel will create a set of evenly distributed classes between the data's minimum and maximum values. The steps for drawing a histogram using the **Data Analysis** tool are shown in the following example.

EXAMPLE   LONG-DISTANCE PHONE BILLS   Chap02\Ex2-5.xls

Referring to the example in Section 2.1, the following represents a sample of 200 Sprint Canada long-distance phone bills (in dollars):

Bills
96.34
94.24
107.00
118.58
66.16
15.37
7.46
67.81
0.00
22.64
3.81
72.66
7.24
100.78
34.58
⋮

(complete data on disk)

Create a histogram for these data.

Setting Up the Template

1. Import the headings and data from file **Chap02\Ex2-5.xls** into Columns A and B. Note that the data to be analyzed and the upper class limits need to be entered on the same sheet from which the **Data Analysis** tool will be used.

2. Select Cells A1:B1 and click **Bold** and **Center** on the toolbar.

3. From the **Tools** menu, choose **Data Analysis**. From the **Data Analysis** tools list, choose **Histogram** and click **OK**. For **Input Range**, enter the location of the data, **A1:A201**. For **Bin Range**, enter the location of the upper class limits, **B1:B9**. Click the check box to specify **Labels,** as variable names have been included in the **Input Range** and **Bin Range**. Select the **Output Range** button then click in the box and enter **D1** as the range. Select the **Chart Output** check box and click **OK** to generate the histogram.

Histogram dialog box

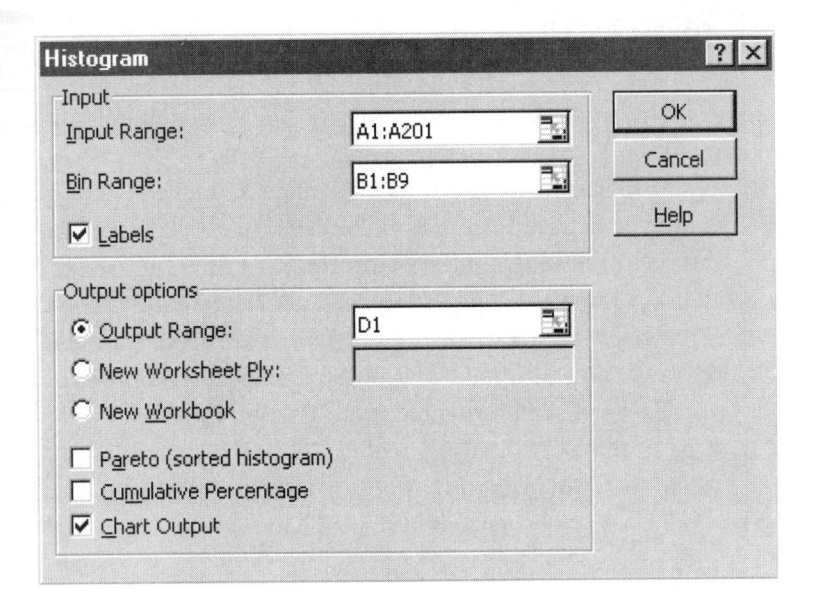

- **Input**
  - Input Range: A1:A201
  - Bin Range: B1:B9
  - ☑ Labels
- **Output options**
  - ⦿ Output Range: D1
  - ○ New Worksheet Ply:
  - ○ New Workbook
  - ☐ Pareto (sorted histogram)
  - ☐ Cumulative Percentage
  - ☑ Chart Output

Buttons: OK, Cancel, Help

4. Click once anywhere on the histogram. When you place the mouse pointer on one of the eight square handles on the border, it changes to a double-arrow pointer. By clicking and dragging on any of these handles, enlarge the chart to improve its appearance. Release the mouse pointer when the desired size is achieved. To move the chart, click anywhere within the chart and drag it to the desired location in the spreadsheet.

Editing/Formatting the Histogram

To edit a chart, click anywhere within the chart box. The frame of the chart should change and the menu at the top of the screen changes to those commands used by charts. All the editing features of bar charts discussed in Section 1.2 (pp. 10–11) can be used with histograms as well.

1. To move a title or label, click on it and drag it to the desired location. To edit a title, click to select it; click the text a second time to edit. Move the cursor to the word **Bin** on the histogram, and click twice. Type **Bills** as the new *x*-axis title.

2. To remove the space between the bars, double-click on the bars. On the **Options** tab, change the value of the **Gap width** to **0** and click **OK**.

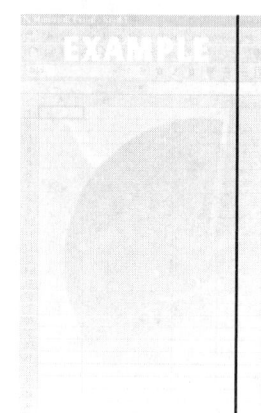

Discussing the Outcome

The frequency distribution indicates that there are 69 bills less than $15, 35 bills between $15 and $30, and so on.

To superimpose the cumulative relative frequency polygon onto the histogram, simply select the **Cumulative Percentage** check box in the histogram dialog box in Step 3 of "Setting Up the Template."

Recall that the **Data Analysis** tools are predefined routines that allow users to generate results that are often extremely difficult or impossible to generate with formulas and functions. The output generated is now part of the spreadsheet and can be edited and formatted just like any other area of a spreadsheet. The disadvantage of the **Data Analysis** tool is that, if the data set changes after the tool has been used, the histogram is not updated to reflect the changes. A new histogram would have to be generated if any changes were made in the data.

PROBLEMS

1. A sales manager has recorded the dollar values, to the nearest $10, of orders obtained by 32 salespeople in the last week. The results are as follows:

450	380	280	330	390	310	340	240
260	370	460	480	320	320	330	410
330	290	280	510	250	230	420	340
380	430	370	360	360	310	390	350

Construct a frequency distribution with six classes, each of width 50. The first class should be "220 to under 270."

2. A controller has recorded the following observations on the random variable "profit per retail sale (in dollars)":

68	38	54	88	62	58	76	34
52	60	70	42	72	48	62	54
50	82	30	52	60	56	42	64
78	50	44	36	64	58	48	52

Construct a histogram with six classes, each of width 10. The first class should be "30 to under 40."

3. The following data show the list prices of a sample of houses for sale advertised in the Montreal *Gazette*. Prices are expressed to the nearest $1,000 (that is, the first

PROBLEMS

two were listed for $108,900 and $129,000 respectively). The cities of Beaconsfield and Brossard were sampled with the following results:

Beaconsfield
108.9 129 139 149 149.9 184.5 189 239 274.9 340 369 525

Construct a histogram with five classes, each of width 85. The first class should be "105 to under 190."

Brossard
79.9 92.5 104 125 128 129 129 135 135 165 229 299 325 329

Construct a histogram with five classes, each of width 50. The first class should be "79.5 to under 129.5."

CHAPTER 3
Tables for Categorical Data

3.1 CREATING ONE-WAY SUMMARY TABLES FOR CATEGORICAL VARIABLES

The **PivotTable** command summarizes information about different variables in tables. The results can be summarized in terms of frequencies, means, and totals.

EXAMPLE **FACULTY STUDY** Chap03\Ex3-1.xls

A small private college is conducting a study of its faculty. Each faculty member's identification number, gender, department number, and rank (1 = sessional lecturer, 2 = visiting professor, 3 = associate professor, 4 = professor) are recorded, as in the following sample:

ID	Gender	Dept	Rank
300	F	7	2
301	M	6	3
302	F	5	4
303	M	2	3
304	F	7	4
305	M	4	4
306	M	2	1
307	F	7	2
308	M	7	3
309	M	4	4
310	M	3	4
311	F	1	2
312	M	8	4
313	M	4	3
⋮	⋮	⋮	⋮
⋮	⋮	⋮	⋮

(complete data on disk)

Prepare a one-way summary table for the "department" categorical variable.

EXAMPLE

Setting Up the Template

1. Import the headings and data from file **Chap03\Ex3-1.xls** into Columns A to D.

2. To format, select Cells A1:D1 and click **Bold** on the toolbar. Select Cells A1:D122 and click **Center** on the toolbar. Providing column headings in the data set allows the **PivotTable Wizard** to form the on-screen buttons for each variable that will be needed to construct tables.

3. From the **Data** menu, choose **PivotTable Report**. Select **Microsoft Excel list or database** and click the **Next** button.

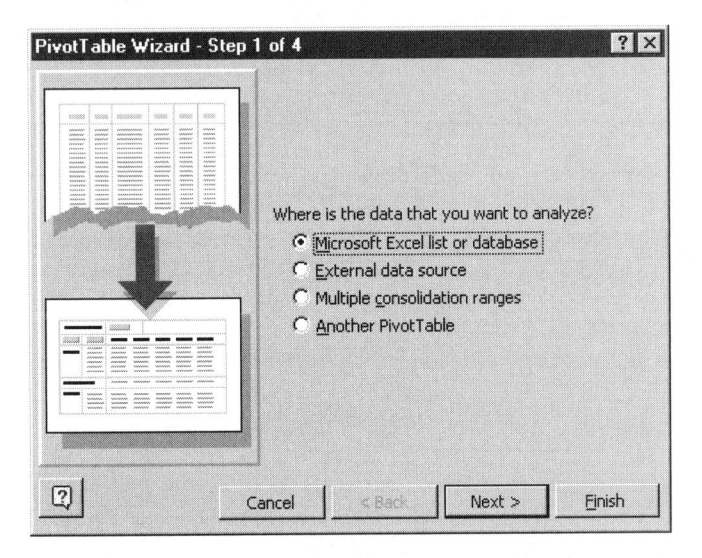

4. Enter **A1:D122** as the **Range** and then click the **Next** button.

5. The third **PivotTable** dialog box allows you to indicate the variable to be summarized and how the data are to be analyzed. Labels corresponding to the column headings for the four variables appear on the right side of the dialog box. Drag the **Dept** label and drop it (by releasing the mouse button) in the box named **Row**. The name "Dept" now appears in the **Row** box. In order to obtain a count of the number of faculty members in each department, drag **Dept** again from the list on the right side and drop it in the **Data** area. Double-click the name in the box titled **Data** and, in the **PivotTable Field** dialog box, select **Summarize by**: **Count** and click **OK**. Click the **Next** button.

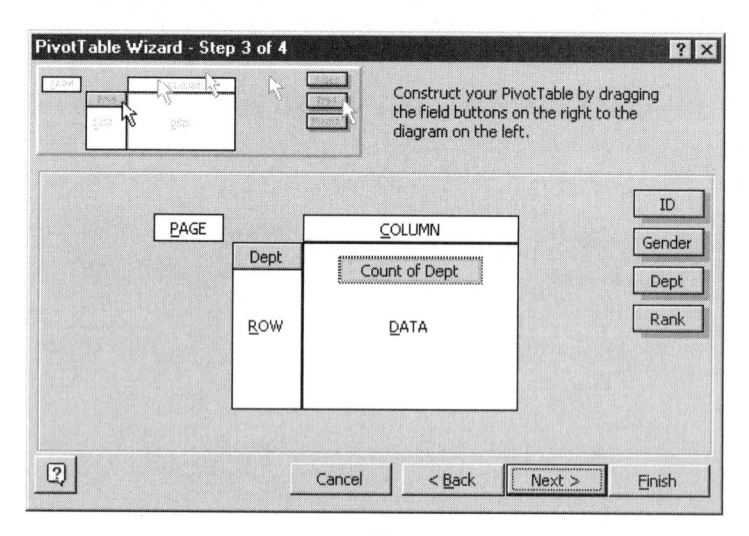

6. The fourth dialog box allows you to place the **PivotTable**. Enter **F5** for **Existing Worksheet** then click the **Options** tab. Enter **Count of Dept** for **Name**. Click the **OK** button, then click the **Finish** button to produce the table.

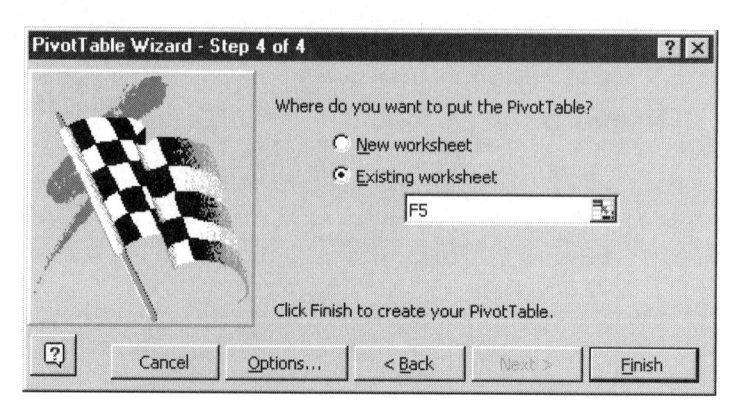

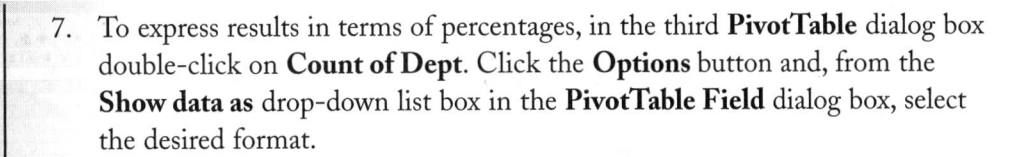

7. To express results in terms of percentages, in the third **PivotTable** dialog box double-click on **Count of Dept**. Click the **Options** button and, from the **Show data as** drop-down list box in the **PivotTable Field** dialog box, select the desired format.

3.2 CREATING TWO-WAY SUMMARY TABLES FOR CATEGORICAL VARIABLES

The **PivotTable** command can be used to create a summary table for two categorical variables.

EXAMPLE FACULTY STUDY

Chap03\Ex3-2.xls

Refer to the example in Section 3.1, which presented data on faculty members at a college:

ID	Gender	Dept	Rank
300	F	7	2
301	M	6	3
302	F	5	4
303	M	2	3
304	F	7	4
305	M	4	4
306	M	2	1
307	F	7	2
308	M	7	3
309	M	4	4
⋮	⋮	⋮	⋮

(complete data on disk)

Construct a two-way summary table for the categorical variables "department" and "gender."

EXAMPLE

Setting Up the Template

1. Import the headings and data from file **Chap03\Ex3-2.xls** into Columns A through D. Repeat Steps 3 and 4 from Section 3.1 (p. 43) to arrive at the third **PivotTable** dialog box.

2. Drag and drop the **Gender** label into the **Row** box, and drag and drop the **Dept** label into the **Column** area. Select the **Dept** label a second time from the list on the right side, and drag and drop it into the **Data** area. Double-click the name in the box titled **Data** and, in the **PivotTable Field** dialog box, select **Summarize by: Count** and click **OK**. Click the **Next** button.

3. In the fourth dialog box, select **Existing Worksheet** and enter **F20**. Click the **Options** button, and enter **Gender by Dept** in the **Name** edit box. Click the **OK** button, then click the **Finish** button to produce the table.

PROBLEMS

Chap03\P3-2#1.xls

1. A study of 80 Canadian homes with personal computers collected data on whether or not the home computer had a CD-ROM drive and whether or not the home computer was running Microsoft Office:

CD-ROM	Office
no	yes
yes	yes
yes	yes
no	no
yes	yes
no	yes
yes	yes
yes	no
no	yes
.	.
.	.
.	.

(complete data on disk)

a. Prepare a one-way summary table for each categorical variable.
b. Prepare a two-way summary table for the data set. Compute the row percents.

SECTION 3.2 • CREATING TWO-WAY SUMMARY TABLES FOR CATEGORICAL VARIABLES

PROBLEMS

Chap03\P3-2#2.xls

2. A study of a company's advertising in 213 Canadian cities and towns considered the advertising medium used (1 = radio, 2 = print, 3 = television) and the advertising agency used (1 = AdCan, 2 = AdPlus):

Medium	Agency
3	2
2	1
1	2
3	1
2	2
2	2
3	2
1	1
.	.
.	.

(complete data on disk)

a. Prepare a one-way summary table for each categorical variable.
b. Prepare a two-way summary table for the data set. Compute the row percents.

Chap03\P3-2#3.xls

3. A study of 132 people from Quebec recorded their gender and whether or not they smoke:

Gender	Smoker
M	no
F	yes
M	no
M	yes
F	no
F	no
F	yes
M	yes
M	no
F	no
.	.
.	.

(complete data on disk)

a. Prepare a one-way summary table for each categorical variable.
b. Prepare a two-way summary table for the data set. Compute the row percents.

PROBLEMS

Chap03\Ex3-1.xls

Chap03\Ex3-1.xls

4. Referring to the example discussed in Section 3.1, construct a two-way summary table for the categorical variables "rank" and "gender."

5. Referring to the example discussed in Section 3.1, construct a two-way summary table for the categorical variables "rank" and "department."

CHAPTER 4
Probability Distributions

4.1 COMPUTING EXPECTED VALUES, VARIANCES, AND STANDARD DEVIATIONS

A *random variable* is a variable whose numerical value is determined by the outcome of a random trial. A *probability distribution* is a listing of the values of a random variable along with their associated probabilities. The probabilities of a probability distribution must sum to 1.

For probability distributions, the *mathematical expectation* or *mean* of the distribution is defined by the sum of the products of all the values of the random variable, weighted by their probabilities:

$$E(X) = \sum X_i P(X_i)$$

The term "mathematical expectation" is used since the expected value is rarely one of the possible values of a random variable but a value close to the average of a large number of observations. The larger the number of observations, the more likely the average will be close to the expected value.

Similarly, we can obtain the following expression for the *variance* of a probability distribution:

$$\sigma^2(X) = \sum X_i^2 P(X_i) - [E(X)]^2$$

The *standard deviation* $\sigma(X)$ is the square root of the variance.

Excel does not have a **Data Analysis** tool to determine the expected value, variance, and standard deviation of a random variable, so Excel formulas and functions must be used instead. The steps for setting up the template are shown in the following example.

EXAMPLE VISITS TO COMPUTER LAB Chap04\Ex4-1.xls

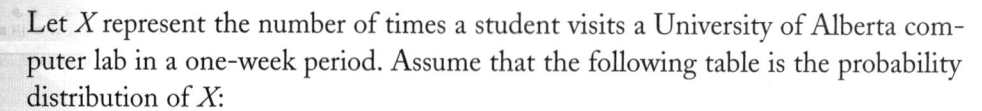

Let X represent the number of times a student visits a University of Alberta computer lab in a one-week period. Assume that the following table is the probability distribution of X:

X	P(X)
0	0.25
1	0.20
2	0.05
3	0.40
4	0.10
Total	**1**

What is the expected value and standard deviation of X?

Setting Up the Template

1. Type the headings and data above in Cells A3:B8 or import the headings and data from Cells A3:B8 of file **Chap04\Ex4-1.xls**.

2. In Cell A1, enter the title **Computing Expected Values, Variances, and Standard Deviations**. In Cell C3, enter the heading **X^2P(X)**. In Cells A9:A12, enter the headings **Total, Expected Value, Variance**, and **Standard Deviation**. Select Cells A1:C1 and, from the **Format** menu, choose **Column, Width**. Change the width to **20**. To format the headings, select Cells A1, A3:C3, and A9:C12 and click **Bold** on the toolbar. Select Cells A3:C3 and click **Center** on the toolbar.

3. In Cell C4, enter the formula **=A4^2*B4**. Copy Cell C4 and paste to Cells C5:C8.

4. In Cell B9, enter the formula **=SUM(B4:B8)** to compute the sum of the probabilities.

5. In Cell C9, enter the formula **=SUM(C4:C8)** to compute the sum $\sum X_i^2 P(X_i)$.

6. In Cell B10, enter the formula **=SUMPRODUCT(A4:A8,B4:B8)** to compute the expected value $E(X) = \sum X_i P(X_i)$.

EXAMPLE

7. In Cell B11, enter the formula **=C9-B10^2** to compute the variance
$$\sigma^2(X) = \Sigma X_i^2 P(X_i) - [E(X)]^2.$$

8. In Cell B12, enter the formula **=SQRT(B11)** to compute the standard deviation by taking the square root of the variance.

Discussing the Outcome

Now that the template has been set up to compute the expected value, variance, and standard deviation for a random variable, we can observe the effects of a change in the probability distribution on the results. For example, change the probabilities in Cells B4:B8 to 0.15, 0.10, 0.30, 0.25, and 0.20, respectively. Notice how the expected value, variance, and standard deviation change to 2.25, 1.69, and 1.299, respectively.

PROBLEMS

1. A company employs salespeople to market its products. They are paid $250 a week plus a 12% commission on the sales they make. The revenue from items a salesperson sells in one week is a random variable whose probability distribution is the same for all salespeople, as follows:

X	1,100	1,200	1,300	1,400	1,500
P(X)	0.35	0.12	0.18	0.20	0.15

   a. Find $E(X)$ and $\sigma^2(X)$.
   b. Find the expected salary and its standard deviation.

2. What is the expected value and variance for the random variable X = number of heads in three tosses of a fair coin?

3. A box contains 6 red marbles and 3 green marbles. One marble is drawn at random from the box and replaced, then a second marble is drawn at random from the box. If both marbles are green, you win $5; if both are red, you lose $1; and if they are different colours, you win or lose nothing. If the random variable X = the amount you win, what is the expectation of X?

PROBLEMS

4. A population consists of 6 Francophones and 4 Anglophones. A random sample of 2 people is selected. What is the mean and variance of the random variable $X =$ the number of Francophones in the sample?

5. The probability distribution for the random variable $X =$ number of equipment breakdowns per day in a manufacturing plant is as follows:

Value of X	0	1	2	3	4
Probability	0.65	0.25	0.05	0.03	0.02

What is the expected number of breakdowns per day and what is the standard deviation?

4.2 COMPUTING BINOMIAL PROBABILITIES

A binomial distribution has the following characteristics:

1. The experiment consists of a fixed number of n repeated trials.
2. Each trial has exactly two possible outcomes: *success* or *failure*.
3. The probability of success p does not change from trial to trial.
4. $P(\text{success}) = p$ and $P(\text{failure}) = 1 - p$.
5. The trials are independent.
6. The binomial random variable X equals the number of successes in n trials.
7. Sampling is *with* replacement from an infinite population (that is, $n/N < 0.05$).

Binomial probabilities can be determined using the formula

$$P(X = x) = C_x^n \, p^x (1 - p)^{(n - x)} \text{ for } X = 1, 2, 3, \ldots, n$$

where C_x^n is the number of different ways of choosing x objects from a total of n objects:

$$C_x^n = \frac{n!}{x!(n - x)!} \text{ where } n! = n(n - 1)(n - 2) \ldots (2)(1)$$

Mean and *variance* can be determined with these formulas:

$$E(X) = np$$

$$\sigma^2(X) = np(1 - p)$$

$$\sigma(X) = \sqrt{np(1 - p)}$$

The shape of the binomial distribution for a fixed sample size will be as follows:

If $p = 0.5$, then the distribution is symmetrical.

If $p < 0.5$, then the distribution is skewed to the right.

If $p > 0.5$, then the distribution is skewed to the left.

The further away from 0.5 that p is, the more pronounced the skewness.

Excel does not have a **Data Analysis** tool to compute binomial probabilities, so Excel formulas and functions are used instead. The steps for setting up the template are shown in the following example.

EXAMPLE

ERRORS ON TAX RETURNS

Chap04\Ex4-2.xls

Revenue Canada reports that 8% of all tax returns contain errors.

a. A random sample of 10 income tax returns is checked by an auditor. Find the probability that she will find errors in 3 of the tax returns.

b. Find the probability that she will find errors in less than 2 of the 10 tax returns that she checks.

c. If the auditor checks a random sample of 10 tax returns, find the probability that at least 1 contains errors.

d. What would be the expected number of tax returns containing errors in a sample of 10 tax returns? What is the variance?

Setting Up the Template

1. In Cell A1, enter the title **Computing Binomial Probabilities**. In Cells A3:E3, enter the headings **n, p, Mean, Variance,** and **Std. Dev.** In Cells A6:F6, enter the headings **X, P(X=x), P(X<=x), P(X<x), P(X>x),** and **P(X>=x)**. Select Cells A1:F1 and, from the **Format** menu, choose **Column, Width**. Change the width to **12**. To format the headings, select Cell A1 and click **Bold** on the toolbar. Select Cells A3:E3 and A6:F6 and click **Bold** and **Center** on the toolbar.

2. In Cell A4, enter the sample size (for this example, **10**), and in Cell B4, enter the probability of success (**0.08**).

3. In Cell C4, enter the formula **=A4*B4** to compute the mean.

4. In Cell D4, enter the formula **=A4*B4*(1-B4)** to compute the variance.

5. In Cell E4, enter the formula **=SQRT(D4)** to compute the standard deviation.

6. In Cell A7:A17, enter the numbers **0, 1, 2, . . ., 10**.

7. In Cell B7, enter the formula **=BINOMDIST(A7,$A$4,$B$4,FALSE)** to

compute $P(X = x)$. A7 = the number of successes, $A$4 = the sample size, $B$4 = the probability of success, and FALSE computes the probability of exactly X successes.

8. In Cell C7, enter the formula **=BINOMDIST(A7,$A$4,$B$4,TRUE)** to compute $P(X \leq x)$. TRUE computes the probability of X or fewer successes.

9. In Cell D7, enter the formula **=C7-B7** to compute $P(X < x)$.

10. In Cell E7, enter the formula **=1-C7** to compute $P(X > x)$.

11. In Cell F7, enter the formula **=1-D7** to compute $P(X \geq x)$.

12. Copy Cells B7:F7 and paste to Cells B8:F17.

Discussing the Outcome

Now that the template has been set up to compute binomial probabilities, we observe that $P(X = 3) = 0.0343$, $P(X < 2) = 0.8121$, $P(X \geq 1) = 0.5656$, $E(X) = 0.8$, and $\sigma^2(X) = 0.736$. We can observe the effects of a change in the probability of success and/or sample size on the results. For example, change the probability of success in Cell B4 to 0.15. Notice how the results change to $P(X = 3) = 0.1298$, $P(X < 2) = 0.5443$, $P(X \geq 1) = 0.8031$, $E(X) = 1.5$, and $\sigma^2(X) = 1.275$.

PROBLEMS

1. A manufacturer will accept a large shipment of a particular component if a random sample of 10 components has no more than 1 defective component. Find the probability that a large shipment will be rejected if the shipment contains exactly 5% defective components.

2. Twenty-five percent of Canadians have type O blood. In a random sample of 20 people, find

    a. the probability that exactly 5 have type O blood
    b. the probability that at least 1 has type O blood
    c. the probability that at most 3 have type O blood
    d. the expected number with type O blood and the standard deviation

PROBLEMS

3. Ninety percent of Canadians are right-handed. In a random sample of 15 people, find the probability that

   a. at least 2 of them are left-handed
   b. at most 3 are left-handed
   c. between 1 and 4, inclusive, are right-handed

4. A family clothing store in Edmonton permits garments to be returned within 14 days of purchase. The store's experience shows a 10% chance of return. If 20 garments are sold to 20 independent individuals, find

   a. the probability that not more than 20% of the garments will be returned
   b. the probability that at least 1 of the garments will be returned
   c. the expected number of returns and the variance

5. It has been estimated that 15% of Canadians never drink coffee. In a survey of a dozen people, find the probability that

   a. less than 3 do not drink coffee
   b. more than 2 do not drink coffee
   c. between 1 and 4, inclusive, are not coffee drinkers

4.3 COMPUTING HYPERGEOMETRIC PROBABILITIES

The *hypergeometric distribution* is very similar to the binomial distribution. It differs from the binomial distribution in the following characteristics:

1. Trials are *not* independent.

2. The probability of success *will* vary from trial to trial.

3. Sampling is *without* replacement from a finite population (that is, $n/N \geq 0.05$).

4. The variable S equals the number of successes in the population.

5. Hypergeometric is binomial but with the population size N.

Hypergeometric probabilities can be determined using the formula

$$P(X = x) = \frac{C_x^S C_{n-x}^{N-S}}{C_n^N}$$

Mean and variance can be determined with these formulas:

$$E(X) = np$$

$$\sigma^2(X) = \frac{N-n}{N-1} \cdot np(1-p)$$

where $p = S/N$ and $\frac{N-n}{N-1}$ is called the *finite population correction factor* (FPCF)

$$\sigma(X) = \sqrt{\frac{N-n}{N-1} \cdot np(1-p)}$$

Excel does not have a **Data Analysis** tool to compute hypergeometric probabilities, so Excel formulas and functions are used instead. The steps for setting up the template are shown in the following example.

EXAMPLE: STUDENT SELECTION FOR COMPLAINTS — Chap04\Ex4-3.xls

A class of 50 students at the University of Alberta consists of 30 men and 20 women. A random sample of 5 students is selected to complain to the dean about the professor's sloppy teaching habits.

a. What is the expected number of women in the sample?

b. What is the standard deviation of the number of women?

c. What is the probability that the sample will contain exactly two women?

Setting Up the Template

1. In Cell A1, enter the title **Computing Hypergeometric Probabilities**. In Cells A3:F3, enter the headings **n, S, N, Mean, Variance**, and **Std. Dev**. In Cells A6:B6, enter the headings **X** and **P(X=x)**. Select Cells A1:F1 and, from the Format menu, choose **Column, Width**. Change the width to **12**. To format the headings, select Cell A1 and click **Bold** on the toolbar. Select Cells A3:F3 and A6:B6 and click **Bold** and **Center** on the toolbar.

2. In Cell A4, enter the sample size (for this example, **5**). In Cell B4, enter the number of successes (**20**). In Cell C4, enter the population size (**50**).

3. In Cell D4, enter the formula **=A4*B4/C4** to compute the mean.

4. In Cell E4, enter the formula **=(C4-A4)/(C4-1)*D4*(1-B4/C4)** to compute the variance.

5. In Cell F4, enter the formula **=SQRT(E4)** to compute the standard deviation.

6. In Cells A7:A12, enter the numbers **0, 1, 2, 3, 4, 5**.

7. In Cell B7, enter the formula **=HYPGEOMDIST(A7,$A$4,$B$4,$C$4)** to compute $P(X = 0)$. A7 = the number of successes, $A$4 = the sample size, $B$4 = the number of successes, and $C$4 = the population size. (The symbol $ makes the address absolute rather than relative.)

8. Copy Cell B7 and paste to Cells B8:B12.

Discussing the Outcome

Now that the template has been set up to compute hypergeometric probabilities, we observe that $E(X) = 2$, $\sigma(X) = 1.05$, and $P(X = 2) = 0.3641$. We can observe the effects of a change in the number of successes, sample size, or population size on the results. For example, change the number of successes in Cell B4 to 32. Notice how the results change to $E(X) = 3.2$, $\sigma(X) = 1.03$, and $P(X = 2) = 0.1910$.

PROBLEMS

1. An exclusive club in Westmount, Quebec, has 200 members, 40 of whom are CEOs. In a random sample of 8 club members, find the probability that exactly 2 are CEOs.

2. Suppose that a TSE broker plans to select 4 out of 8 new stock issues and that, unknown to the broker, 3 of the 8 will result in substantial profits and 5 will result in losses. What is the probability that less than 2 of the profitable issues will appear in the broker's selection?

3. A shelf has 16 bottles of Molson Export and 4 bottles of Labatt 50 in a random arrangement. If a bartender takes 5 bottles of beer off the shelf, find the probability that he gets exactly 2 bottles of Molson Export and 3 bottles of Labatt 50.

4. In a given semester, 120 students are registered in a University of Western Ontario statistics class in a given semester. Of these students, 25% have a PC at home. The university's M.I.S. department is conducting a study on personal computing and therefore selects a sample of 10 students to fill out a questionnaire. What is the variance of the number of students in the sample who have a PC at home?

5. A police department in Halifax has 20 police officers who are eligible for promotion. Of these officers, 12 were born in Halifax. Supposing that only 5 of the police officers are chosen for promotion, what is the probability that exactly 1 of the 5 promoted officers was born in Halifax?

4.4 COMPUTING POISSON PROBABILITIES

A *Poisson distribution* has the following characteristics:

1. The Poisson random variable X equals the number of occurrences in a given interval of time or space.

2. The number of occurrences in any interval is independent of the number of occurrences in any other interval.

3. There is no theoretical limit on the number of occurrences in an interval, although a relatively large number is unlikely.

4. The mean (λ) equals the mean number of occurrences in a given interval of time or space.

5. The variance equals λ.

6. The mean number of occurrences in an interval is proportional to the size of the interval (that is, $\lambda = 30$ per hour implies that $\lambda = 15$ per 30 minutes or $\lambda = 60$ per 2 hours).

7. For any infinitesimally small portion of the interval, the probability of 2 or more occurrences is negligible.

Poisson probabilities can be determined with the formula

$$P(X = x) = \lambda^x e^{-\lambda} / x!, \text{ for } X = 1, 2, 3, \ldots, \infty$$

where e is the base of the natural logarithms (≈ 2.71828).

Poisson distributions are right-skewed for all values of λ; however, as λ increases, the distribution becomes more and more bell shaped.

Excel does not have a **Data Analysis** tool to compute Poisson probabilities, so Excel formulas and functions are used instead. The steps for setting up the template are shown in the following example.

EXAMPLE — TRAFFIC FLOW   Chap04\Ex4-4.xls

On an uninterrupted stretch of the Trans-Canada Highway, the traffic flow in one direction averages 120 cars per hour. What is the probability of

a. no cars passing in 1 minute?

b. more than 3 cars passing in 1 minute?

c. fewer than 2 cars passing in 1 minute?

EXAMPLE

Setting Up the Template

1. In Cell A1, enter the title **Computing Poisson Probabilities**. In Cell A3, enter the heading **Lambda**. In Cells A5:F5, enter the headings **X, P(X=x), P(X<=x), P(X<x), P(X>x)**, and **P(X>=x)**. Select Cells A1:F1 and, from the **Format** menu, choose **Column, Width.** Change the width to **12**. To format the headings, select Cell A1 and click **Bold** on the toolbar. Select Cells A3 and A5:F5 and click **Bold** and **Center** on the toolbar.

2. In Cell B3, enter the value for lambda (for this example, 120/60 = **2**).

3. In Cells A6:A26, enter the numbers **0, 1, 2, . . ., 20**.

4. In Cell B6, enter the formula **=POISSON(A6,$B$3,FALSE)** to compute $P(X = x)$. A6 = the number of successes, $B$3 = lambda, and FALSE computes the probability of exactly X successes.

5. In Cell C6, enter the formula **=POISSON(A6,$B$3,TRUE)** to compute $P(X \leq x)$. TRUE computes the probability of X or fewer successes.

6. In Cell D6, enter the formula **=C6-B6** to compute $P(X < x)$.

7. In Cell E6, enter the formula **=1-C6** to compute $P(X > x)$.

8. In Cell F6, enter the formula **=1-D6** to compute $P(X \geq x)$.

9. Copy Cells B6:F6 and paste to Cells B7:F26.

Discussing the Outcome

Now that the template has been set up to compute Poisson probabilities, we observe that $P(X = 0) = 0.1353$, $P(X > 3) = 0.1429$, and $P(X < 2) = 0.4060$. We can observe the effects of a change in the value of lambda on the results. For example, change the value of lambda in Cell B3 to 2.5. Notice how the results change to $P(X = 0) = 0.0821$, $P(X > 3) = 0.2424$, and $P(X < 2) = 0.2873$.

PROBLEMS

1. The number of tax returns with an arithmetical error processed at a regional tax office in London, Ontario, arrive with a mean of 5 per hour, following a Poisson distribution. Find the probability that

   a. more than 4 tax returns with errors are processed in 30 minutes
   b. at most 2 tax returns with errors are processed in 10 minutes
   c. between 6 and 8 tax returns with errors, inclusive, are processed in 2 hours

2. A bakery receives an average of 6 requests per 8-hour day for a particular type of pie. Assume that the number of daily requests follows a Poisson distribution. Find

   a. the probability that exactly 4 requests are made between 1:00 p.m. and 3:00 p.m. on a particular day
   b. the probability that at least 12 requests are made over a period of two 8-hour days
   c. the variance of the number of requests per 8-hour day for a particular type of pie
   d. the standard deviation

3. The mean number of arrivals at a branch of the Royal Bank in Saint John, New Brunswick, between 10:30 a.m. and 11:30 a.m. is 30, and the arrivals follow a Poisson distribution. Find the probability that

   a. exactly 2 people will arrive in 10 minutes
   b. at most 5 people will arrive in 30 minutes
   c. at least 1 person will arrive in 5 minutes

4. Suppose that, in a carpet manufacturing operation, an average of 2 flaws occur per 10 metres of material, and the flaws follow a Poisson distribution. Find the probability that

   a. a given 10-metre segment will have 1 or fewer flaws
   b. a given 20-metre segment will have at least 2 flaws
   c. a given 30-metre segment will have no flaws

5. The *Globe and Mail* reports an average of 36 burglary attempts per year in a small Ontario town. Determine the probability that

   a. there are 4 or more burglary attempts next month
   b. there are exactly 2 burglary attempts next month

COMPUTING NORMAL PROBABILITIES

A normal distribution has the following characteristics:

1. The distribution has two parameters: the mean (μ) and the standard deviation (σ).

2. The mean will determine where the curve is centred on the x-axis, and the standard deviation will determine how spread out the curve is. The larger the standard deviation, the flatter the curve.

3. It is symmetrical about its mean and is bell shaped, regardless of μ and σ.

4. The mean, median, and mode are all equal.

5. The total area under the curve is equal to 1, 50% of the area falling to the left of the mean and 50% of the area falling to the right of the mean.

6. The probability of a value falling between point a and point b is equal to the area under the curve bounded by point a and point b.

7. Each different combination of μ and σ specifies a different normal distribution.

8. Regardless of the shape (μ and σ) of the normal distribution, the following statements hold:

    The probability of a normal random variable being within $\pm 1\sigma$ is 0.6826.

    The probability of a normal random variable being within $\pm 2\sigma$ is 0.9544.

    The probability of a normal random variable being within $\pm 3\sigma$ is 0.9974.

9. A normal random variable can range from $-\infty$ to ∞, but values a large distance from the mean occur rarely. The curve therefore only approaches but never actually reaches the x-axis (it is *asymptotic*).

10. The area between the mean and any point above the mean equals the area between the mean and the same point below the mean.

Excel does not have a **Data Analysis** tool to compute normal probabilities, so Excel formulas and functions are used instead. The steps for setting up the template are shown in the following example.

EXAMPLE

STATISTICS GRADES

Chap04\Ex4-5.xls

Grades in a statistics course at Montreal's Concordia University are normally distributed with a mean of 70 and a standard deviation of 10.

a. Find the probability that a randomly selected student gets a grade under 80.

b. Find the probability that a randomly selected student gets a grade over 75.

c. Find the probability that a randomly selected student gets a grade between 80 and 90.

d. If the professor decides to fail 10% of the students, what grade would a student have to obtain to pass the course?

Setting Up the Template

1. In Cell A1, enter the title **Computing Normal Probabilities**. Select Cells A1:B1, and from the **Format** menu, choose **Column, Width**. Change the width to **25**. Type or import the headings from Cells A3:A4, A6:A7, A10:A11, A14:A15, A17, and A20:A23 of file **Chap04\Ex4-5.xls**. To format the headings, select Cells A1:A23 and click **Bold** on the toolbar.

2. To reference the *X*-values on which to compute the probabilities, enter the following templates:

   Cell A8— ="P(X<="&B7&")"

   Cell A12— ="P(X>="&B11&")"

   Cell A16— ="P(X<="&B15&")"

   Cell A18— ="P(X<="&B17&")"

   Cell A19— ="P("&B15&"<=X<="&B17&")"

3. In Cell B3, enter the value for the mean (for this example, **70**). In Cell B4, enter the value for the standard deviation (**10**).

4. In Cell B7, enter the *X* value for the lower tail probability (**80**).

5. In Cell B8, enter the formula **=NORMDIST(B7,B3,B4,TRUE)** to compute the lower tail probability. B7 = the X value, B3 = the mean, B4 = the standard deviation, and TRUE computes the cumulative probability associated with X.

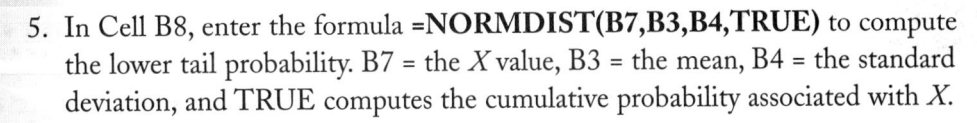

6. In Cell B11, enter the X value for the upper tail probability (**75**).

7. In Cell B12, enter the formula **=1-NORMDIST(B11,B3,B4,TRUE)** to compute the upper tail probability.

8. In Cell B15, enter the lower X value for the interval probability (**80**).

9. In Cell B16, enter the formula **=NORMDIST(B15,B3,B4,TRUE)** to compute $P(X \leq x)$.

10. In Cell B17, enter the upper X value for the interval probability (**90**).

11. In Cell B18, enter the formula **=NORMDIST(B17,B3,B4,TRUE)** to compute $P(X \leq x)$.

12. In Cell B19, enter the formula **=ABS(B18-B8)** to compute the interval probability.

13. In Cell B22, enter the cumulative probability (**0.1**).

14. In Cell B23, enter the formula **=NORMINV(B22,B3,B4)** to compute the X value associated with the cumulative probability.

Discussing the Outcome

Now that the template has been set up to compute normal probabilities, we observe that $P(X < 80) = 0.8413$, $P(X > 75) = 0.3085$, $P(80 < X < 90) = 0.1359$, and the grade the student would have to obtain to pass the course is 57.18. We can observe the effects of a change in the mean and standard deviation on the results. For example, change the value of the mean in Cell B3 to 72 and the standard deviation in Cell B4 to 5. Notice how the results change to $P(X < 80) = 0.9452$, $P(X > 75) = 0.2743$, $P(80 < X < 90) = 0.0546$, and the grade the student would have to obtain to pass the course is 65.59.

PROBLEMS

1. The number of hours taken by truck drivers to haul a load from Montreal to Vancouver is normally distributed with a mean of 70 hours and a standard deviation of 5 hours.

   a. If a trucking company officially allows 72 hours for the Montreal–Vancouver run, what percentage of drivers arrive in Vancouver late?
   b. What is the probability that a randomly selected run will take between 65 and 72.5 hours?
   c. The trucking company gives a bonus to the fastest 5% of drivers. Within what amount of time must a driver complete the run in order to qualify for the bonus?

2. A statistical analysis of long-distance telephone calls made from the headquarters of Canadian Marconi indicates that the length of these calls is normally distributed with $\mu = 240$ seconds and $\sigma = 40$ seconds.

   a. What percentage of these calls lasted less than 180 seconds?
   b. What is the length of a particular call if only 1% of all calls are shorter?

3. The mean height of MUC (Montreal Urban Community) police officers is 71 inches with a standard deviation of 2 inches. The heights of MUC police officers are normally distributed.

   a. Find the probability that a randomly selected police officer is between 70 and 71.5 inches tall.
   b. Find the probability that a randomly selected police officer is more than 72 inches tall.
   c. Find the probability that a randomly selected police officer is less than 70.8 inches tall.

4. The length of sardines arriving at a Maritime cannery is believed to be normally distributed with a mean of 4.6 centimetres and a standard deviation of 0.2 centimetres.

   a. What percentage of all sardines are longer than 5 centimetres?
   b. What percentage of all sardines are between 4.35 and 4.85 centimetres long?

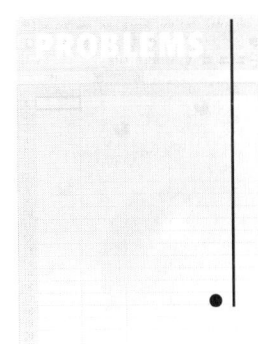

5. Air Canada knows from experience that the number of suitcases it loses each week on the Montreal–Toronto route is a normally distributed random variable with a mean of 31.4 and a standard deviation of 4.5. What is the probability that in any given week it will lose

  a. at most 20 suitcases?
  b. less than 30 suitcases?
  c. between 25 and 35 suitcases, inclusive?

4.6 COMPUTING EXPONENTIAL PROBABILITIES

An *exponential distribution* had the following characteristics:

1. The exponential random variable X equals the time between two consecutive occurrences of an event.

2. $\mu = 1/\lambda$ and $\sigma = 1/\lambda$.

3. The exponential distribution is closely related to the Poisson distribution, as they use the same parameter λ = mean number of occurrences in a given interval of time.

4. All exponential distributions are right-skewed.

Exponential probabilities can be determined with the following formulas:

$$P(X \leq a) = 1 - e^{-\lambda a}, \text{ for } X > 0 \text{ and } \lambda > 0$$

$$P(X \geq a) = e^{-\lambda a}$$

$$P(a \leq X \leq b) = P(X \leq b) - P(X \leq a) = e^{-\lambda a} - e^{-\lambda b}$$

Excel does not have a **Data Analysis** tool to compute exponential probabilities, so Excel formulas and functions are used instead. The steps for setting up the template are shown in the following example.

EXAMPLE

DRIVE-IN TELLER SERVICE

Chap04\Ex4-6.xls

The length of time required to serve customers arriving at a Royal Bank drive-in teller's window is known to be exponentially distributed with a mean of 5 minutes. What is the probability that the time between 2 consecutive customers being served will be at least 1 minute?

Setting Up the Template

1. In Cell A1, enter the title **Computing Exponential Probabilities**. In Cells A3:A7, enter the headings **Mean, Lambda, Value of X, P(X<=x)**, and **P(X>=x)**. Select cells A1:B1 and, from the **Format** menu, choose **Column, Width**. Change the width to **12**. To format the headings, select Cells A1:A7 and click **Bold** on the toolbar.

2. In Cell B3, enter the mean of the exponential distribution (in this example, **5**).

3. In Cell B4, enter the formula **=1/B3** to compute the corresponding value of λ.

4. In Cell B5, enter the value of the random variable X (**1**).

5. In Cell B6, enter the formula **=EXPONDIST(B4,B5,TRUE)** to compute $P(X \leq x)$. TRUE computes the cumulative probability associated with X.

6. In Cell B7, enter the formula **=1-B6** to compute $P(X \geq x)$.

Discussing the Outcome

Now that the template has been set up to compute exponential probabilities, we observe that $P(X \geq 1) = 0.8187$. We can observe the effects of a change in the value of the mean on the resulting probability. For example, change the value of the mean in Cell B3 to 4. Notice how the probability now changes to $P(X \geq 1) = 0.7788$.

PROBLEMS

1. Emergency calls are received at a 911 exchange in Toronto following an exponential distribution. The mean time between successive calls is 4 minutes.

    a. If a call is received at 9:00 a.m., find the probability that the next call will not be received until after 9:20 a.m.
    b. If a call is received at 9:00 a.m., find the probability that the next call will not be received until after 9:15 a.m.

2. The lifetime of an electric bulb is an exponentially distributed random variable with a mean of 200 hours. What is the probability that such a bulb will last at most 50 hours?

3. Tax returns with an arithmetical error arrive at a regional tax office in Sudbury, Ontario, at a mean of 5 per hour, following a Poisson distribution. If a tax return with an error is processed at 10:00 a.m., find the probability that the next return with an error will be received before 10:20 a.m.

4. A drop-in information service run by a Manitoba government agency has an average of 10 clients per hour on a typical morning. Clients arrive according to a Poisson distribution.

    a. What is the mean time between arrivals?
    b. What is the probability that the time between 2 consecutive arrivals will be more than 8 minutes?
    c. What is the probability that the time between 2 consecutive arrivals will be at least 3 but at most 5 minutes?

5. A small business has a printer that experiences a failure of some type an average of once every 2 hours of operation, and failures follow a Poisson distribution.

    a. What is the probability that the time between 2 successive failures will be less than 3 hours?
    b. What is the probability that the time between 2 successive failures will be between 1 hour and 4 hours, inclusive?

CHAPTER 5
Sampling and Simulating Sampling Distributions

5.1 SELECTING A RANDOM SAMPLE FROM A POPULATION

A random sample is one that is drawn in such a way that each member of the population has an equally likely chance of being selected for the sample.

The **Data Analysis: Sampling** tool can be used to select a random sample from a population (with replacement) provided the population values are already entered on a spreadsheet. The steps for using the **Data Analysis: Sampling** tool are shown in the following example.

EXAMPLE **LONG-DISTANCE PHONE BILLS** Chap05\Ex5-1.xls

Suppose the following represents a population of 200 Sprint Canada long-distance phone bills for a particular household:

Bills
96.34
94.24
107.00
118.58
66.16
15.37
7.46
67.81
0.00
.
.
.

(complete data on disk)

Select a random sample of 30 bills from this population.

Setting Up the Template

1. Import the headings and data from Column A of file **Chap05\Ex5-1.xls**. In Cell B1, enter the heading **Sample**. To format the headings, select Cells A1:B1 and click **Bold** and **Center** on the toolbar.

2. From the **Tools** menu, choose **Data Analysis**. From the **Data Analysis** tools list, choose **Sampling** and click **OK**. For **Input Range**, enter **A2:A201**. Click the **Random** button and, for **Number of Samples**, enter **30**. Click the **Output Range** button then click in the box and enter **B2**. Click **OK** to generate the random sample.

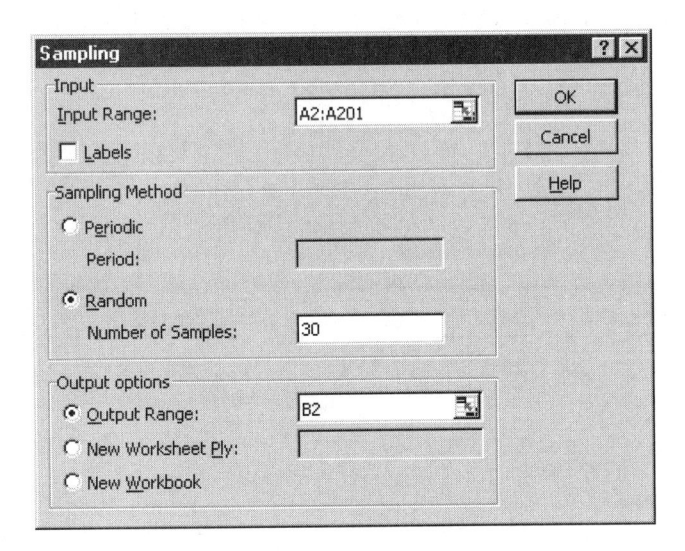

Discussing the Outcome

The random sample now appears in Cells B2:B31. Had the population of values not already been entered on a spreadsheet, the RANDBETWEEN function could be used to select a random sample of a given size. For example, in Cell D2, enter the following: =RANDBETWEEN(1,200). Copy Cell D2 and paste to Cells D3:D31. Notice that Cells D2:D31 now contain integer random numbers between 1 and 200. The phone bills on the population list that correspond to the random numbers would then be selected as members for the sample. As with the **Data Analysis: Sampling** tool, the random numbers are selected with replacement, so it is possible to observe random numbers repeating. To obtain a random sample

EXAMPLE

without replacement, reselect the sample until none of the random numbers repeat themselves by hitting the F9 key to recalculate functions in the spreadsheet.

Recall that the output of a **Data Analysis** tool exists only as numbers and not functions in the spreadsheet. If random numbers repeat themselves, the **Data Analysis: Sampling** tool must be applied until no numbers appear more than once if a random sample without replacement is desired.

PROBLEMS

Chap05\Ex5-1.xls

1. Using the Data Analysis: Sampling tool, select a simple random sample of size 42 from the population of Sprint Canada long-distance phone bills in the example above.

Chap05\P5-1#2.xls

2. Use the RANDBETWEEN function to generate a random sample of size 10 from a list of companies in the Canadian fast-food sector:

Company
McDonald's Restaurants of Canada Ltd.
Cara Operations Ltd.
KFC Canada
The TDL Group Ltd.
Scott's Hospitality Inc.
Pizza Hut Canada
Versa Services
A&W Food Services of Canada
Subway Franchise Systems of Canada Ltd.
Dairy Queen Canada Inc.
Burger King Restaurants of Canada Inc.
Wendy's Restaurants of Canada Inc.
⋮
(complete data on disk)

Which companies become members of the sample?

5.2 USING RANDOM NUMBERS TO SIMULATE THE CENTRAL LIMIT THEOREM

The *central limit theorem* states that the sample mean is approximately normal regardless of the underlying population provided that the sample size is "sufficiently" large ($n \geq 30$). As the sample size is increased, the sampling distribution of the sample mean will more closely approach the normal distribution.

$$E(\bar{x}) = \mu$$

$$\sigma(\bar{x}) = \sigma/\sqrt{n}$$

The standard deviation (error) of the sample mean is less than the population standard deviation since there is less of a chance that a sample mean will take on an extreme value than there is that a single value from the population will take on an extreme value.

The **Data Analysis: Random Number Generation** tool can be used to generate random numbers selected from several distributions, including the normal and binomial. The steps for using random numbers to simulate the central limit theorem are shown in the following example.

EXAMPLE VARIATION IN PIPE LENGTHS Chap05\Ex5-2.xls

There is variation in a production process where the actual lengths of pipes produced vary uniformly between 19.5 and 20.5 centimeters. At regular intervals a random sample of 5 pipes is selected and the mean length is recorded. After 600 samples have been selected, the distribution of the sample means is plotted.

Setting Up the Template

1. In Cells A1:F1, enter the headings **Sample #1, Sample #2, Sample #3, Sample #4, Sample #5**, and **Sample Mean**. Select Cells A1:F1 and, from the **Format** menu, choose **Column, Width**. Change the width to **15**. To format the headings, select Cells A1:F1 and click **Bold** and **Center** on the toolbar.

2. From the **Tools** menu, choose **Data Analysis**. From the **Data Analysis** tools list, choose **Random Number Generation** and click **OK**. For **Number of Variables**, enter **5**, and for **Number of Random Numbers**, enter **600**. Choose **Uniform** from the **Distribution** drop-down list; under **Parameters**, enter **Between 19.5 and 20.5**. Click the **Output Range** button then click in the box and enter **A2**. Click **OK** to generate the random numbers.

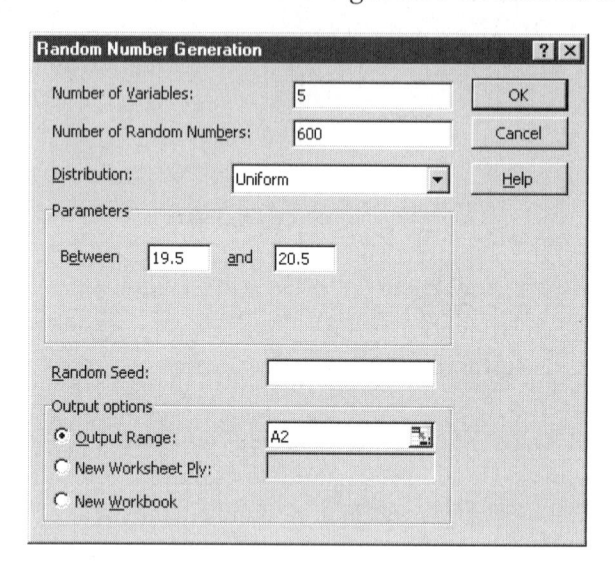

3. In Cell F2, enter the formula =**AVERAGE(A2:E2)**. Copy Cell F2 and paste to Cells F3:F601.

4. In Cell E603, enter the label **Grand Mean**, and in Cell E604, enter the label **Standard Error**.

5. In Cell F603, enter the formula =**AVERAGE(F2:F601)**. In our example, we get 20.0117, which is a good approximation of $\mu = (19.5 + 20.5) / 2 = 20$.

6. In Cell F604, enter the formula =**STDEV(F2:F601)**. In our example, we get 0.1347, which is a good approximation of σ. Note that $\sigma = \sqrt{(20.5 - 19.5)^2/12}$ and $\sigma(\bar{x}) = \sigma/\sqrt{5} = 0.129$.

7. To format, select Cells E603:F604 and click **Bold** on the toolbar.

8. From the **Tools** menu, choose **Data Analysis**. From the **Data Analysis** tools list, choose **Histogram** and click **OK**. For **Input Range**, enter the location of the data, **F1:F601**. Click the check box to specify **Labels**, as the variable name has been included in the input range. Click the **Output Range** button, then click the box and enter **I1**. Select the **Chart Output** check box, and click **OK** to generate the histogram.

SECTION 5.2 • USING RANDOM NUMBERS TO SIMULATE THE CENTRAL LIMIT THEOREM

EXAMPLE

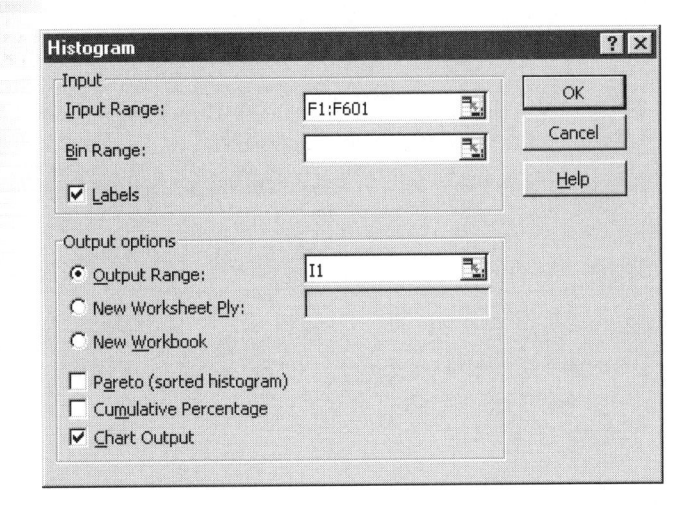

9. Click once anywhere within the chart box. When you place the mouse pointer on one of the eight square handles on the border, it changes to a double-arrow pointer. By clicking and dragging on any of these handles, enlarge the chart to improve its appearance. Release the mouse pointer when the desired size is achieved. To move the chart, click anywhere within the chart and drag it to the desired location in the spreadsheet.

10. Move the cursor to the word "Bin" on the histogram, and click twice. Type the new *x*-axis label **Sample Mean**.

The histogram can be further edited as described in Section 2.5 (p. 39).

Discussing the Outcome

Notice the bell shape of the histogram. Even though the population values are uniformly distributed and the sample size is small, the sampling distribution of the sample mean is approximately normal.

Since the **Data Analysis: Random Number Generation** tool can be used to generate random numbers selected from several different distributions, including the normal and binomial, we can observe the effects of changes in the sampling distribution of the sample mean. For a given sample size, we can explore the effect of the shape of the population (binomial, Poisson, etc.) on the sampling distribution of the sample mean. In addition, we can investigate the effect of increasing the sample size on the shape of the sampling distribution of the sample mean. For example, a smoother, more bell-shaped distribution than the histogram generated in the above example can be obtained by using 1000 or 2000 samples of size 5 instead of only 600.

PROBLEMS

1. Demonstrate the central limit theorem by creating a worksheet that will display 100 values from a Poisson distribution, with a mean of 2.5, for each of 45 samples. Compute the mean for each sample and show that

$$E(\bar{x}) = \mu \text{ and } \sigma(\bar{x}) = \sigma/\sqrt{n}$$

Use a histogram to plot the distribution of the sample means to emphasize the point that, even though the population is not normal (in this case, it follows a Poisson distribution), the sampling distribution of the sample mean is approximately normal.

2. Demonstrate the central limit theorem by creating a worksheet that will display 1000 values from a discrete uniform distribution ranging from 1 to 6, inclusive, intended to represent the rolling of a fair die for each of 30 samples. Compute the mean for each sample and show that

$$E(\bar{x}) = \mu \text{ and } \sigma(\bar{x}) = \sigma/\sqrt{n}$$

Use a histogram to plot the distribution of the sample means to emphasize the point that, even though the population is not normal (in this case, it follows a discrete uniform distribution), the sampling distribution of the sample mean is approximately normal.

CHAPTER 6
Estimation

6.1 CONFIDENCE INTERVAL FOR THE POPULATION MEAN USING z

An *interval estimate* includes a range of possible values for which there is an associated degree of confidence that it will include the value of the population parameter of interest. One of the limitations of point estimates is that they rarely result in a value that exactly matches the population parameter; therefore, it is not possible to indicate how much confidence or accuracy can be placed in the estimate. Interval estimates, on the other hand, are constructed to provide information concerning the precision of the estimate since they are accompanied by a measure of confidence. A point estimate is placed at the centre of the interval, the length of which depends on the degree of confidence and the standard error.

A confidence interval for the population mean using the z-distribution can be determined with the following formula:

$$\mu = \bar{x} \pm z(1-\alpha/2) \cdot \sigma / \sqrt{n}$$

The z-distribution can be used in the following situations:

1. If the sample size is greater than or equal to 30, the central limit theorem applies and the sample mean is approximately normal regardless of whether or not the population variance is known or whether or not the population is normally distributed. The sample standard deviation provides an adequate estimate for the population standard deviation if it is not known.

2. If the sample size is less than 30, the population is normally distributed, and the population variance is known, then the sample mean follows a normal distribution.

When raw data are not available, Excel formulas and functions can be used instead. To construct the confidence interval for the population mean, we must have the sample size, sample mean, standard deviation, and confidence level, which can be entered as values, formulas, or references. If the sample size, sample mean, and standard deviation need to be computed from raw data, then the **COUNT**, **AVERAGE**, and **STDEV** functions, respectively, can be used. Alternatively, the **Descriptive Statistics** option of the **Data Analysis** tool can be used to compute the half-width of the confidence interval by selecting the **Confidence Interval for Mean** check box in the **Descriptive Statistics** dialog box. The steps for setting up the confidence interval are shown in the following example.

EXAMPLE: EXPENDITURES ON ACADEMIC SUPPLIES   Chap06\Ex6-1.xls

A population of 1225 students registered in a faculty at Queen's University in Kingston, Ontario, is to be sampled to estimate the mean amount spent on academic supplies in the university bookstore in a semester. A random sample of 100 students spent a mean amount of $440 with a standard deviation of $65. Establish a 95% confidence interval for the mean amount spent by these students.

Setting Up the Template

1. Type or import the labels in Column A of file **Chap06\Ex6-1.xls**. Select one cell in Column A. From the **Format** menu, choose **Column**, **Width**. Change the width to **30**. To format the title, select Cells A1:A2 and click **Bold** on the toolbar.

2. In Cell B4, enter the sample size (for this example, **100**).

3. In Cell B5, enter the sample mean (**440**).

4. In Cell B6, enter the standard deviation (**65**).

5. In Cell B7, enter the desired confidence level (**0.95**).

6. In Cell B9, enter the formula **=B6/SQRT(B4)** to compute the standard deviation of the sample mean.

7. In Cell B10, enter the formula **=NORMSINV(0.5+(B7/2))** to compute the z-value.

EXAMPLE

8. In Cell B11, enter the formula **=B9*B10** to compute the half-width of the interval by multiplying the z-value by the standard deviation of the sample mean.

9. In Cell B13, enter the formula **=B5-B11** to compute the lower bound of the interval by subtracting the half-width from the sample mean.

10. In Cell B14, enter the formula **=B5+B11** to compute the upper bound of the interval by adding the half-width to the sample mean.

Complete steps 11-15 only if population size is known:

11. In Cell B16, enter the population size (for this example, **1125**).

12. In Cell B17, enter the formula **=SQRT((B16-B4)/(B16-1))** to compute the finite population correction factor (FPCF).

13. In Cell B18, enter the formula **=B11*B17** to compute the new half-width by multiplying the old half-width in Cell B11 by the FPCF.

14. In Cell B19, enter the formula **=B5-B18** to compute the lower bound of the interval by subtracting the half-width from the sample mean.

15. In Cell B20, enter the formula **=B5+B18** to compute the upper bound of the interval by adding the half-width to the sample mean.

Discussing the Outcome

Probabilistic Interpretation of a Confidence Interval

In repeated sampling from a normal population, $100(1 - \alpha)$% of all intervals that are constructed in a similar manner from simple random samples of a given size will, in the long run, include the population parameter. One particular interval, however, will either correctly include the population parameter or incorrectly exclude it.

EXAMPLE

Practical Interpretation of a Confidence Interval

We are $100(1 - \alpha)$% confident that the interval will contain the actual value of the population parameter. In our example, we are 95% confident that the mean amount spent on academic supplies in the university bookstore is between $427.79 and $452.21.

Now that the template has been set up, we can observe the effect of changes in the sample size, sample mean, standard deviation, and level of confidence on the width of the confidence interval. For example, to determine the effect of a change in the confidence level on the width of the confidence interval, change the value in Cell B6 from 0.95 to 0.99. Notice how the z-value in Cell B10 changes to 2.576 and the confidence interval bounds in Cells B13 and B14 widen to $423.95 and $456.05, respectively. All other things being equal, for a fixed sample size, the more confidence used, the wider (less precise) the interval. The implication is that the more certain we wish to be that the interval contains the population parameter, the wider the interval will have to be.

On the other hand, as the sample size increases, both confidence and precision can increase. For example, change the confidence level in Cell B6 from 0.95 to 0.99 and the sample size in Cell B4 from 100 to 200. Notice how the confidence interval bounds in Cells B13 and B14 narrow to $429.17 and $450.83, respectively. The confidence and precision are now higher than in the original confidence interval of $427.79 to $452.21.

PROBLEMS

1. A computer store manager wants to estimate the mean price of modems with 95% confidence. A random sample of 25 modems has a mean price of $94.50. The standard deviation of the price of modems is known from past experience to be $20. Construct a 95% confidence interval for the mean price of modems, assuming that the price is a normally distributed random variable.

2. A sales manager wants to estimate the mean profit per sale. A random sample of 225 sales are found to have a mean profit of $527 with a standard deviation of $75. Construct a 90% confidence interval for the true mean profit per sale.

3. A broker on the Toronto Stock Exchange wishes to investigate the amount of time between the placement and execution of a market order. She samples 32 orders and finds that the mean time to execution is 23.8 minutes with a standard deviation of 3.16 minutes. Construct a 95% confidence interval for the mean time to execution.

PROBLEMS

4. An auditor has examined 9 accounts and determined that the mean balance is $4,565. She knows from past experience that the standard deviation of the population of accounts is $1,110. Construct a 99% confidence interval estimate for the population mean, assuming that the population of accounts is normally distributed.

5. Revenue Canada is conducting an audit of the 11,308 outlets of a large convenience store chain. It is interested in determining the average error in reported net income last year for all outlets in the chain. The size of the chain is too large for a census, so 1,000 outlets are randomly selected and audited. The audits show the sample mean error in reported income for a given outlet to be $12,985 and the sample standard deviation to be $4,025. Construct a 95% confidence interval for the mean error in reported income per outlet.

6.2 CONFIDENCE INTERVAL FOR THE POPULATION MEAN USING t

A confidence interval for the population mean using the t-distribution can be determined with the following formula:

$$\mu = \bar{x} \pm t(1 - \alpha/2;\ n - 1) \cdot s/\sqrt{n}$$

The t-distribution can be used in the following situation: If the sample size is less than 30, the population is normally distributed and the population variance is unknown, then the sample mean follows a t-distribution with $n - 1$ degrees of freedom.

When the population standard deviation is unknown, we use the sample standard deviation (s) in place of the population standard deviation (σ). The sample standard deviation is not a good estimator of the population standard deviation when the sample size is small, so the t-distribution compensates for this fact. Small samples are necessary in situations where samples are destructive or certain traits exist only in small amounts.

The t-distribution has the following characteristics:

1. It has the same basic bell shape as the normal distribution.

2. It has a mean of 0 and is symmetrical about its mean.

3. It is really a family of distributions rather than just a single distribution, since there is a distribution for each degrees of freedom $(n - 1)$ value.

4. In general, the *t*-distribution tends to be flatter and more spread out than the standard normal distribution.

5. The *t*-distribution approaches the standard normal distribution as the sample size (and thereby the degrees of freedom) increases.

When raw data are not available, Excel formulas and functions can be used to construct a confidence interval estimate for the population mean. To construct such an interval, we must have the sample size, sample mean, sample standard deviation, and confidence level. If the sample size, sample mean, and standard deviation need to be computed from raw data, then the **COUNT**, **AVERAGE**, and **STDEV** functions, respectively, can be used. Alternatively, the **Descriptive Statistics** option of the **Data Analysis** tool can be used to compute the half-width of the confidence interval by selecting the **Confidence Interval for Mean** check box in the **Descriptive Statistics** dialog box. The sample size, sample mean, and sample standard deviation can also be entered as values when raw data are not available. The steps for setting up the confidence interval are shown in the following example.

EXAMPLE: WEIGHT OF LUMBER SHIPMENTS — Chap06\Ex6-2.xls

British Columbia lumber is sold by weight, which is a normally distributed random variable. A random sample of 6 shipments from the Canadian Lumber Corporation yielded the following weights (in kilograms):

| 687 | 811 | 602 | 789 | 650 | 781 |

Find a 90% confidence interval for the mean weight per shipment from the corporation.

Setting Up the Template

1. Change the name of **Sheet1** to **Data** by double-clicking on the name at the bottom of the sheet and typing the new one. Change the name of **Sheet2** to **Example**. Select the **Data** sheet and enter the title **Weight** in Cell A1 and the data in Cells A2:A7. Select the title in Cell A1 and click **Bold** and **Center** on the toolbar.

SECTION 6.2 • CONFIDENCE INTERVAL FOR THE POPULATION MEAN USING t

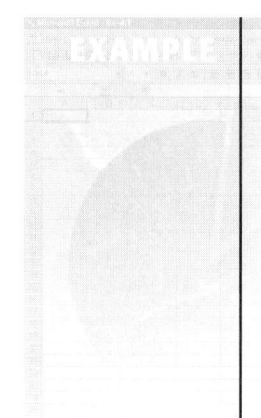

2. Select the **Example** sheet. Import or type the labels from Column A of file **Chap06\Ex6-2.xls**. Select one cell in Column A. From the **Format** menu, choose **Column, Width**. Change the width to **30**. To format the title, select Cells A1:A2 and click **Bold** on the toolbar.

3. In Cell B4, enter the formula **=COUNT(Data!$A$2:$A$7)** to compute the sample size. (**Data!** refers to the sheet titled "Data.")

4. In Cell B5, enter the formula **=AVERAGE(Data!$A$2:$A$7)** to compute the sample mean.

5. In Cell B6, enter the formula **=STDEV(Data!$A$2:$A$33)** to compute the sample standard deviation.

6. In Cell B7, enter the desired confidence level (for this example, **0.90**).

7. In Cell B9, enter the formula **=B6/SQRT(B4)** to compute the estimated standard deviation of the sample mean.

8. In Cell B10, enter the formula **=TINV(1-B7,B4-1)** to compute the t-value.

9. In Cell B11, enter the formula **=B9*B10** to compute the half-width of the interval by multiplying the t-value by the estimated standard deviation of the sample mean.

10. In Cell B13, enter the formula **=B5-B11** to compute the lower bound of the interval by subtracting the half-width from the sample mean.

11. In Cell B14, enter the formula **=B5+B11** to compute the upper bound of the interval by adding the half-width to the sample mean.

Discussing the Outcome

In our example, we are 90% confident that the mean weight per shipment is between 649.54 and 790.46 kilograms. Now that the template has been set up, we can observe the effect of changes in the sample size, sample mean, standard deviation, and level of confidence on the width of the confidence interval. We can also observe the effect of a change in one of the data values on the confidence interval. For example, change the observation in Cell A7 on the **Data** sheet from 781 to 881. On the **Example** sheet, notice how the sample mean in Cell B5 changes from 720 to 736.67 and the sample standard deviation in Cell B6 changes from 85.65 to 106.97. The confidence interval bounds in Cells B13 and B14 widen to 648.67 and 824.66 kilograms, respectively.

PROBLEMS

1. A random sample of 9 cars were crashed into a wall at 30 kilometres per hour. The mean cost of repairs was $1,638 with a standard deviation of $321. Assuming that repair costs are normally distributed, construct a 99% confidence interval for the mean cost of repairing cars involved in head-on collisions at 30 kilometres per hour.

2. The monthly incomes of a random sample of 5 students are $225, $405, $210, $340, and $370. Assuming that income is normally distributed, find the 99% confidence interval estimate for the mean income of the student population.

3. The time required to train an employee to use a new software package is normally distributed. A manager has recorded the time taken to train a random sample of 15 employees and found the mean time to be 6 days with a standard deviation of 2 days. Construct a 95% confidence interval estimate for mean time required to train employees to use the new software package.

4. Assume the weights of male students at the University of British Columbia are normally distributed. Construct a 90% confidence interval for the average weight of male students if a random sample of 25 male students had an average weight of 152.82 pounds with a standard deviation of 17.09 pounds.

5. A manufacturer of alarm systems makes an alarm that is sensitive to movement. The quality-control department tests a random sample of 16 alarms to determine the amount of movement required for the alarm to activate. The coded results are as follows:

| 3 | 8 | 9 | 8 | 9 | 5 | 9 | 6 |
| 6 | 5 | 2 | 3 | 8 | 7 | 6 | 8 |

Establish a 95% confidence interval for the population mean.

6.3 CONFIDENCE INTERVAL FOR THE POPULATION PROPORTION

The sample proportion ($\bar{p}$) is approximately normal for "large" values of n [that is, $np \geq 5$ and $n(1 - p) \geq 5$].

$$p = \bar{p} \pm z(1 - \alpha/2) \sqrt{\frac{\bar{p}(1 - \bar{p})}{n - 1}}$$

In the long run, $100(1 - \alpha)\%$ of the intervals constructed in a similar manner will include the population proportion. We are therefore $100(1 - \alpha)\%$ confident that the interval actually contains the population proportion. Use the finite population correction factor (FPCF) in $\sigma(\bar{p})$ if $n/N \geq 5$

Excel does not have a **Data Analysis** tool to construct confidence intervals for the population proportion, so Excel formulas and functions can be used instead. To construct the confidence interval for the population proportion, we must have the sample size, number of successes in the sample, and confidence level. These can be entered as values or, if the sample size and number of successes must be obtained from raw data, the **PivotTable Wizard** can be used and the results copied into the appropriate cells. The steps for setting up the confidence interval are shown in the following example.

EXAMPLE — SAME-SEX BENEFITS

Chap06\Ex6-3.xls

A sample of 100 voters in Ontario contained 64 people who favoured same-sex benefits in the Canadian workplace. Between what limits can we be 95% confident that the proportion of voters in the province who favour same-sex benefits is contained?

Setting Up the Template

1. Type or import the labels in from Column A in file **Chap06\Ex6-3.xls**. Select one cell in Column A. From the **Format** menu, choose **Column, Width**. Change the width to **30**. To format the title, select Cells A1:A2 and click **Bold** on the toolbar.

2. In Cell B4, enter the sample size (in this example, **100**).

3. In Cell B5, enter the number of successes (**64**).

4. In Cell B6, enter the desired confidence level (**0.95**).

5. In Cell B8, enter the formula **=B5/B4** to compute the sample proportion.

6. In Cell B9, enter the formula =**NORMSINV(0.5+B6/2)** to compute the z-value.

7. In Cell B10, enter the formula =**SQRT((B8*(1-B8))/(B4-1))** to compute the standard deviation of the sample proportion.

8. In Cell B11, enter the formula =**B9*B10** to compute the half-width of the interval by multiplying the z-value by the standard deviation of the sample proportion.

9. In Cell B13, enter the formula =**B8-B11** to compute the lower bound of the interval by subtracting the half-width from the sample proportion.

10. In Cell B14, enter the formula =**B8+B11** to compute the upper bound of the interval by adding the half-width to the sample proportion.

Discussing the Outcome

We are 95% confident that the proportion of voters in the province who favour same-sex benefits is between 54.5% and 73.5%. Now that the template has been set up, we can observe the effect on the confidence interval of changes in the sample size, number of successes in the sample, and level of confidence. For example, to determine the effect of a change in the confidence level on the confidence interval, change the value in Cell B6 from 0.95 to 0.99. Notice how the z-value in Cell B9 changes to 2.576 and the confidence interval bounds in Cells B13 and B14 widen to 51.6% and 76.4%, respectively.

PROBLEMS

1. A sample of 125 prescriptions was found to have 31 that allow repeat orders. Construct a 99% confidence interval for the proportion of all prescriptions that allow repeat orders.

PROBLEMS

2. A random sample of 140 computer chips produced by a certain company was found to have 7 defectives. Construct a 95% confidence interval for the proportion of all computer chips made by the company that are defective.

3. A random sample of 160 doctors in Quebec was found to have 88 who oppose user fees for Medicare services. Construct a 90% confidence interval for the proportion of doctors in Quebec who oppose user fees.

4. A random sample of 1200 Quebec voters had 552 who believe that, after a "Yes" vote in a sovereignty referendum, Quebec's current borders would not necessarily be guaranteed. Construct a 96% confidence interval for the population proportion.

5. A random sample of 240 customers of Sprint Canada was found to have 10% who pay their bills through Sprint's new Internet billing service. Construct a 98% confidence interval for the proportion of Sprint Canada customers who pay their bills over the Internet.

6.4 DETERMINING THE SAMPLE SIZE FOR THE POPULATION MEAN

Suppose that you are required to construct a confidence interval with a given level of confidence $1 - \alpha$ and precision or half-width h. Letting $z = z(1 - \alpha/2)$, the formula for the required sample size is

$$n = z^2 \sigma^2 / h^2$$

In practice, the population variance is often unknown and must be estimated. The variance can be estimated in three ways:

1. σ^2 is known from historical data or previous studies.

2. Use a *pilot sample* drawn from the population of interest to compute s^2, and then use s^2 as a *planning value* for σ^2.

3. *Worst-case analysis:* Determine the range (largest value − smallest value) of possible values in the population of interest, and divide the range by 6 to estimate the population variance. This gives the largest possible value for the variance and thus the sample size. It is better to overestimate the sample size than underestimate it.

Be sure always to round the sample size to the next largest integer to assure the desired level of accuracy.

Excel does not have a **Data Analysis** tool to determine sample size, so Excel formulas and functions can be used instead. To determine the sample size for the mean, we must have the confidence level, standard deviation, and desired level of precision. The steps for setting up the test of hypothesis are shown in the following example.

EXAMPLE

COST OF PRODUCING TRANSISTORS

Chap06\Ex6-4.xls

A production manager wants to estimate the mean cost of producing transistors with 99% confidence that the error of the estimate does not exceed $0.50. If the population standard deviation is known from previous experience to be $12, find the sample size necessary to construct the required confidence interval.

Setting Up the Template

1. Type or import the labels from Column A in file **Chap06\Ex6-4.xls**. Select one cell in Column A. From the **Format** menu, choose **Column, Width**. Change the width to **30**. To format the title, select Cells A1:A2 and click **Bold** on the toolbar.

2. In Cell B4, enter the desired confidence level (in this example, **0.99**).

3. In Cell B5, enter the planning value for the population standard deviation (**12**).

4. In Cell B6, enter the desired half-width (error) (**0.50**).

5. In Cell B8, enter the formula **=NORMSINV(0.5+(B4/2))** to compute the necessary z-value.

6. In Cell B9, enter the formula **=((B8*B5)/B6)^2** to compute the sample size.

7. In Cell B10, enter the formula **=ROUNDUP(B9,0)** to round up the sample size generated in Cell B9 to the next largest integer.

SECTION 6.4 • DETERMINING THE SAMPLE SIZE FOR THE POPULATION MEAN

EXAMPLE

Discussing the Outcome

The sample size necessary to construct a confidence interval with 99% confidence that the error of the estimate does not exceed $0.50 is 3822. Now that the template has been set up, we can observe the effect on the sample size of changes in the standard deviation, precision, and level of confidence. For example, to determine the effect of a change in the level of confidence on the sample size, change the level of confidence in Cell B4 from 0.99 to 0.90. Notice how the z-value in Cell B8 changes to 1.645 and the sample size in Cell B10 changes to 1559. By using less confidence, we thereby reduce the necessary sample size. The same effect can be achieved by settling for a wider interval (using less precision).

PROBLEMS

1. A random sample of 20 prescriptions for a particular medication had a mean cost of $22.78 with a standard deviation of $4.19. The cost of prescriptions is known to be approximately normally distributed. Determine the sample size that should be used in order to construct a 99% confidence interval with a total width of $1.00. Assume that the standard deviation of $4.19 obtained from the above sample represents a reasonable planning value for the population standard deviation.

2. A random sample is to be taken with the purpose of determining the mean daily waste disposal of an industrial firm. If a precision of no more than 0.80 tonnes and a 95% confidence level are desired, how large a sample should be taken? Assume the population standard deviation is 4 tonnes and that the variable is normally distributed.

3. The owner of a boutique wishes to estimate the mean profit per sale with 99% confidence and a maximum interval width of $3. The results of a pilot study of 35 sales had a mean of $45 and a variance of 36. Determine the minimum sample size required to construct the specified interval.

4. A credit manager has chosen a sample size to estimate the mean balance of credit accounts to within $0.50. The level of confidence of the interval is 90%. A pilot study of 50 accounts had a mean of $1,219 and a variance of 100. Determine the sample size.

5. The Royal Bank of Canada is considering doing a study to estimate the average number of transactions on savings accounts. A pilot study indicated that the standard deviation is about 2.8 transactions. If the Royal Bank desires a tolerable error of no more than 0.30 transactions and 90% confidence, how large a sample is required?

6.5 DETERMINING THE SAMPLE SIZE FOR THE POPULATION PROPORTION

Suppose you are required to construct a confidence interval with a given level of confidence $1 - \alpha$ and precision or half-width h. Letting $z = z(1 - \alpha/2)$, the formula for the required sample size is

$$n = z^2 p(1 - p)/h^2$$

In practice, the population proportion (p) is often unknown and must be estimated. The population proportion can be estimated in three ways:

1. The population proportion is known from historical data or previous studies.

2. Use a *pilot sample* drawn from the population of interest to compute the sample proportion, and then use it as a *planning value* for the population proportion.

3. *Worst-case analysis:* Use $p = 0.5$ to give the largest possible sample size. It is better to overestimate the sample size than underestimate it.

Be sure always to round the sample size to the next largest integer to assure the desired level of accuracy.

Excel does not have a **Data Analysis** tool to determine sample size, so Excel formulas and functions can be used instead. To determine the sample size for the population proportion, we must have the confidence level, planning value for the population proportion, and desired level of precision. The steps for setting up the sample size are shown in the following example.

EXAMPLE — CAPITAL PUNISHMENT

Chap06\Ex6-5.xls

A random sample of 200 Canadians had 108 who favour the return of capital punishment. A federal government official wants to estimate the proportion of the population in favour of the return of capital punishment with 99% confidence that the estimate is correct to within 3%. Determine the required sample size.

Setting Up the Template

1. Type or import the labels from Column A of file **Chap06\Ex6-5.xls**. Select one cell in Column A. From the **Format** menu, choose **Column, Width**. Change the width to **30**. To format the title, select Cells A1:A2 and click **Bold** on the toolbar.

2. In Cell B4, enter the desired confidence level (for this example, **0.99**).

3. In Cell B5, enter the planning value for the population proportion (**0.54**, which comes from p = 108/200).

4. In Cell B6, enter the desired half-width (error) (**0.03**).

5. In Cell B8, enter the formula **=NORMSINV(0.5+(B4/2))** to compute the necessary z-value.

6. In Cell B9, enter the formula **=(B8/B6)^2*B5*(1-B5)** to compute the sample size.

7. In Cell B10, enter the formula **=ROUNDUP(B9,0)** to round up the sample size generated in Cell B9 to the next largest integer.

Discussing the Outcome

The sample size necessary to estimate the proportion of the population in favour of the return of capital punishment with 99% confidence that the estimate is correct to within 3% is 1832. Now that the template has been set up, we can observe the effect on the sample size of changes in the planning value for the population proportion, precision, and level of confidence. For example, to determine the effect of a change in the desired precision on the sample size, change the precision in Cell B6 from 0.03 to 0.05. Notice how the sample size in Cell B10 changes to 660. By settling for a wider interval (using less precision), we thereby reduce the necessary sample size. The same effect can be achieved through reducing the confidence level.

PROBLEMS

1. Canadian Tire is interested in estimating the proportion of their tires that will become flat within the first 1200 kilometres. Each tire sampled is mounted on a different make of car to assure randomness. One hundred tires are sampled, and 6 of them become flat within 1200 kilometres. What is the sample size required to determine a 95% confidence interval for the population proportion with a half-width of 1.5%?

2. If a pollster is required to construct a 95% confidence interval with a width of 0.08 for the proportion of Canadians in favour of free trade with Mexico, what sample size would be necessary?

3. What sample size is required to estimate the proportion of University of British Columbia students who like statistics with 90% confidence that the error of the estimate does not exceed 0.02? Historical evidence indicates that approximately 5% of UBC students like statistics.

4. A random sample of 120 light bulbs was found to have 9 defectives. Consider this sample as a pilot sample for a quality control manager who wishes to estimate the population proportion of defective light bulbs. She wants the estimate to be within 3% of the true population proportion, and she insists on 98% confidence. How many units would she have to sample?

5. How many people would have to be chosen in order to estimate the proportion of the population of Quebec in favour of sovereignty with 99% confidence that the estimate is within 4% of the actual population proportion?

CHAPTER 7
Hypothesis Testing: One Population

7.1 CREATING A HYPOTHESIS TEST FOR ONE POPULATION MEAN (z-TEST)

In testing a hypothesis for a population mean, the z-test can be used in the following situations:

1. If the sample size is greater than or equal to 30, then the central limit theorem applies and the test statistic is approximately normal regardless of whether or not the population variance is known or whether or not the population is normally distributed.

2. If the sample size is less than 30, the population is normally distributed, and the population variance is known, then the test statistic follows a normal distribution.

Excel does not have a **Data Analysis** tool to test a hypothesis for one population mean, so Excel formulas and functions can be used instead. The decision rule for the test is based on the p-value approach. For each of the three types of hypothesis tests (upper-tailed, lower-tailed, and two-tailed), the decision rule is the following:

$$\text{Do not reject } H_0 \text{ if } p\text{-value} \geq \alpha.$$
$$\text{Reject } H_0 \text{ if } p\text{-value} < \alpha.$$

To test the hypothesis for the mean, we must have the sample size, sample mean, standard deviation, level of significance, and hypothesized value of the mean in the null hypothesis. These can be entered as values, formulas, or references. For example, if the sample size, sample mean, and standard deviation need to be computed, then the **COUNT, AVERAGE,** and **STDEV** functions, respectively, could be used in the appropriate cells. The steps for setting up the hypothesis test are shown in the following example.

EXAMPLE

CONTAINER FILLING WEIGHTS

Chap07\Ex7-1.xls

A production line operates with a filling weight standard of 16 grams per container. Overfilling or underfilling is a serious problem, and the production line should be shut down if this occurs. The standard deviation is known to be 0.8 grams. A quality control inspector samples 40 items every hour and at that time makes the decision of whether or not to shut down the production line for adjustment purposes. If a sample mean of 16.32 grams is observed, what action should be recommended? Use a 5% level of significance.

Setting Up the Template

1. Type or import the labels from Column A of file **Chap07\Ex7-1.xls**. Select one cell in Column A. From the **Format** menu, choose **Column, Width**. Change the width to **30**. To format the headings, select Cells A1, A11, A17, and A22 and click **Bold** on the toolbar.

2. In Cells B3, B4, and B5, enter the sample size, sample mean, and standard deviation respectively. Using the above example, enter **40** in Cell B3, **16.32** in Cell B4, and **0.8** in Cell B5.

3. In Cell B6, enter the formula **=B5/SQRT(B3)** to compute the standard error of the mean.

4. In Cell B7, enter the desired hypothesized value (**16**). In Cell B8, enter the level of significance (**0.05**).

5. In Cell B9, enter the formula **=(B4-B7)/B6** to compute the test statistic.

6. In Cell B12, enter the formula **=NORMSINV(1-B8/2)** to generate the upper critical value for the two-tailed hypothesis test.

7. In Cell B13, enter the formula **=NORMSINV(B8/2)** to generate the lower critical value for the two-tailed hypothesis test.

8. In Cell B14, enter the formula **=2*(1-NORMSDIST(ABS(B9)))** to compute the *p*-value for the two-tailed hypothesis test.

9. The decision rule involves a comparison of the *p*-value and the level of significance. In Cell B15, enter the formula **=IF(B14<B8,"Alternative","Null")**. If the value in Cell B14 (*p*-value) is less than the value in Cell B8 (alpha), then **Alternative** will be displayed in the cell; otherwise, **Null** will be displayed.

10. In Cell B18, enter the formula **=NORMSINV(1-B8)** to generate the critical value for the upper-tailed hypothesis test.

11. In Cell B19, enter the formula **=1-NORMSDIST(B9)** to compute the *p*-value for the upper-tailed hypothesis test.

12. The decision rule for an upper-tailed test follows the same logic as that of the two-tailed test. Enter the formula **=IF(B19<B8,"Alternative","Null")** in Cell B20.

13. In Cell B23, enter the formula **=NORMSINV(B8)** to generate the critical value for the lower-tailed hypothesis test.

14. In Cell B24, enter the formula **=NORMSDIST(B9)** to compute the *p*-value for the lower-tailed hypothesis test.

15. The decision rule for a lower-tailed test follows the same logic as that of the two-tailed test. Enter the formula **=IF(B24<B8,"Alternative","Null")** in Cell B25.

Discussing the Outcome

In the two-tailed test, since 0.0114 < 0.05, the null hypothesis is rejected and **Alternative** is displayed in Cell B15. The production line must therefore be shut down for adjustment purposes.

Now that the template has been set up, we can observe the effect of changes in the sample size, sample mean, standard deviation, and level of significance on the hypothesis test. For example, to determine the effect of a change in the value of the standard deviation on the test statistic, change the value of the standard deviation in Cell B5 from 0.8 to 1.2. Notice how the test statistic changes to 1.69, the two-tailed *p*-value becomes 0.0917, and the null hypothesis is *not* rejected.

PROBLEMS

1. The manager of a small loans company claims that the mean amount deposited by Canadian pensioners per month is at least $750. In a study, a random sample of 36 pensioners had a mean monthly deposit of $720 with a standard deviation of $120. Test the manager's claim at the 5% level of significance.

2. The Quebec manufacturer of an industrial pipe is interested in testing the hypothesis that the mean diameter of the pipes is 12.75 centimetres. For a random sample of 100 pipes, the diameters were found to have a sample mean of 12.73 centimetres, with a sample standard deviation of 0.05 centimetres. Test the null hypothesis that the mean diameter of the pipes is 12.75 centimetres, using a 5% significance level.

3. A criminologist claims that the mean age of people convicted of first-degree murder in Manitoba is less than 27 years of age. A random sample of 64 people convicted of first-degree murder had a mean age of 25.8 and a standard deviation of 3 years. Do these data provide sufficient evidence to support the criminologist's claim at the 5% level of significance?

4. A travel agent claims that the mean cost of a flight from Vancouver to Charlottetown is under $650. A random sample of 49 airline tickets had a mean cost of $630 with a standard deviation of $56. Test the appropriate hypothesis using a 5% level of significance.

5. The purchasing director for an Ontario industrial parts factory is investigating the possibility of purchasing a new type of milling machine. She has determined that the new machine will be bought if there is evidence that the parts produced from the new machine have a higher mean breaking strength than those from the old machine. The process standard deviation of breaking strength is 10 kilograms. A sample of 100 parts taken from the new machine indicated a sample mean breaking strength of 72 kilograms. Past records indicate that the old machine operated with an average breaking strength of 69 kilograms. Use a 10% level of significance. Is there evidence that the purchasing director should buy the new machine?

7.2 CREATING A HYPOTHESIS TEST FOR ONE POPULATION MEAN (*t*-TEST)

In testing a hypothesis for a population mean, the *t*-test can be used in the following situation:

> If the sample size is less than 30, the population is normally distributed, and the population variance is unknown, then the test statistic follows a *t*-distribution with $n - 1$ degrees of freedom.

Excel does not have a **Data Analysis** tool to test a hypothesis for one population mean, so Excel formulas and functions can be used instead. The decision rule for the test is based on the *p*-value approach. For each of the three types of hypothesis tests (upper-tailed, lower-tailed, and two-tailed), the decision rule is the following:

> Do not reject H_0 if *p*-value $\geq \alpha$.
> Reject H_0 if *p*-value $< \alpha$.

To test the hypothesis for the mean, we must have the sample size, sample mean, sample standard deviation, level of significance, and hypothesized value of the mean in the null hypothesis. These can be entered as values, formulas, or references. For example, if the sample size, sample mean, and standard deviation need to be computed, then the **COUNT**, **AVERAGE**, and **STDEV** functions, respectively, could be used in the appropriate cells. The steps for setting up the hypothesis test are shown in the following example.

EXAMPLE — COMMUTING TIME

Chap07\Ex7-2.xls

The personnel department of a Calgary company would like to study the amount of time it takes employees to travel to work in the morning. A random sample of 14 employees is selected, and the average commuting time is found to be 48 minutes with a standard deviation of 20 minutes. At the 1% level of significance, is there evidence that the average morning travel time of employees is 60 minutes? Assume that the times for employees to travel to work are normally distributed.

Setting Up the Template

1. Type or import the labels from Column A of file **Chap07\Ex7-2.xls**. Select one cell in Column A. From the **Format** menu, choose **Column, Width**. Change the width to **30**. To format the headings, select Cells A1, A12, A18, and A23 and click **Bold** on the toolbar.

2. In Cells B3, B4, and B5, enter the sample size, sample mean, and sample standard deviation respectively. Using the above example, enter **14** in Cell B3, **48** in Cell B4, and **20** in Cell B5.

3. In Cell B6, enter the formula **=B5/SQRT(B3)** to compute the standard error of the mean.

4. In Cell B7, enter the desired hypothesized value (**60**). In Cell B8, enter the level of significance (**0.01**).

5. In Cell B9, enter the formula **=(B4-B7)/B6** to compute the test statistic.

6. In Cell B10, enter the formula **=B3-1** to compute the degrees of freedom.

7. In Cell B13, enter the formula **=TINV(B8,B10)** to generate the upper critical value for the two-tailed hypothesis test.

8. In Cell B14, enter the formula **=-(TINV(B8,B10))** to generate the lower critical value for the two-tailed hypothesis test.

9. In Cell B15, enter the formula **=TDIST(ABS(B9),B10,2)** to compute the p-value for the two-tailed hypothesis test.

10. The decision rule involves a comparison of the p-value and the level of significance. In Cell B16, enter the formula **=IF(B15<B8,"Alternative","Null")**. If the value in Cell B15 (p-value) is less than the value in Cell B8 (alpha), then **Alternative** will be displayed in the cell; otherwise, **Null** will be displayed.

11. Type or import the labels in Cells D18:D20 of file **Chap07\Ex7-2.xls**. Select one cell in Column D. From the **Format** menu, choose **Column, Width**. Change the width to **20**. To format the heading, select Cell D18 and click **Bold** on the toolbar.

12. In Cell E19, enter the formula **=TDIST(ABS(B9),B10,1)** to calculate one of the possible p-values.

13. In Cell E20, enter the formula **=1-E19** to calculate the other, complementary, possible *p*-value.

14. In Cell B19, enter the formula **=(TINV(2*B8,B10))** to generate the critical value for the upper-tailed hypothesis test.

15. In Cell B20, enter the formula **=IF(B9<0,E20,E19)** to compute the *p*-value for the upper-tailed hypothesis test.

16. The decision rule for an upper-tailed test follows the same logic as that of the two-tailed test. Enter the formula **=IF(B20<B8,"Alternative","Null")** in Cell B21.

17. In Cell B24, enter the formula **=-(TINV(2*B8,B10))** to generate the critical value for the lower-tailed hypothesis test.

18. In Cell B25, enter the formula **=IF(B9<0,E19,E20)** to compute the *p*-value for the lower-tailed hypothesis test.

19. The decision rule for a lower-tailed test follows the same logic as that of the two-tailed test. Enter the formula **=IF(B25<B8,"Alternative","Null")** in Cell B26.

Discussing the Outcome

In all three tests, the null hypothesis is not rejected and **Null** is displayed in Cells B16, B21, and B26. The sample results therefore support the claim that the average morning travel time of employees is 60 minutes.

Now that the template has been set up, we can observe the effect of changes in the sample size, sample mean, standard deviation, and level of significance on the hypothesis test. For example, to determine the effect of a change in the level of significance on the conclusion of the test, change the value in Cell B8 from 0.01 to 0.05. Notice how the critical values for the two-tailed hypothesis test change to ±2.16 and the null hypothesis is rejected.

PROBLEMS

1. A toy manufacturer is considering the purchase of a new machine to replace an old one. The old machine produces on average 10 units of a given toy per hour. A random sample of 16 hours of production has been examined. The sample mean was 11 units per hour and the sample standard deviation was 2 units. Test the hypothesis that the mean production of the new machine is greater than 10 units per hour. Assume that the distribution of units produced is normal. Use a 5% level of significance.

2. A company is contemplating a switch from manual to automatic handling of a certain product. Before making a decision, the company wants to determine if the mean length of time it takes employees to complete the job manually is less than 1 hour. If not, the new machine will be purchased. A sample of 20 random observations showed that the sample mean was 0.9 hours and the sample standard deviation was 0.3 hours. Using $\alpha = 1\%$, determine what the company should do, based on the sample results. Assume that the time required to complete the job is normally distributed.

3. A real estate agent claims that the average selling price of all homes sold in Montreal in 1997 was $175,000. A random sample of 16 homes had a mean selling price of $171,743 with a standard deviation of $36,400. Test the hypothesis that the mean selling price for all Montreal homes in 1997 was $175,000. Use $\alpha = 5\%$ and assume a normal distribution.

4. The closing price of the stock of a company listed on the Toronto Stock Exchange (TSE) has been recorded for a period of 10 weeks:

| 28.49 | 32.19 | 31.84 | 30.37 | 32.73 | 30.11 | 31.52 | 33.96 | 33.25 | 33.78 |

These prices reflect the volatility of the stock during a successful takeover bid. A stockbroker claims the mean price was greater than $31.00. At the 5% level of significance, would the sample data tend to support the stockbroker's claim?

5. An administrator from Toronto's Mount Sinai Hospital claims that the mean duration of a hospital stay after open-heart surgery is at least 14 days. Duration of hospital stays is assumed to be normally distributed. A random sample of 16 patients showed a mean stay of 12 days with a standard deviation of 1.5 days. At the 1% level of significance, perform the appropriate hypothesis test.

7.3 CREATING A HYPOTHESIS TEST FOR ONE POPULATION PROPORTION

Testing a hypothesis concerning a population proportion is nearly the same as testing a hypothesis concerning a population mean. One of the differences in the procedure includes the use of the population proportion in the formulation of the null and alternative hypotheses. The parameter of interest is the proportion of elements that have a given characteristic. There are also some corresponding changes in the calculation of the test statistic. To compute the standard deviation of the sample proportion, we require knowledge of the population proportion. To allow for the normal approximation (that is, to use the z-distribution), both np and $n(1-p)$ must be at least 5.

Excel does not have a **Data Analysis** tool to test a hypothesis for one population proportion, so Excel formulas and functions can be used instead. The decision rule for the test is based on the p-value approach. For each of the three types of hypothesis tests (upper-tailed, lower-tailed, and two-tailed), the decision rule is the following:

$$\text{Do not reject } H_0 \text{ if } p\text{-value} \geq \alpha.$$
$$\text{Reject } H_0 \text{ if } p\text{-value} < \alpha.$$

To test the hypothesis for the proportion, we must have the sample size, number of successes in the sample, level of significance, and hypothesized value of the proportion in the null hypothesis. These can be entered as values or, if the sample size and number of successes must be obtained from raw data, the **PivotTable Wizard** could be used and the results copied into the appropriate cells. The steps for setting up the hypothesis test are shown in the following example.

EXAMPLE — ERRORS ON TAX RETURNS

Chap07\Ex7-3.xls

In a sample of 500 tax returns, an auditor for Revenue Canada has found 15 that contain computational errors. Does this finding support the claim that 2% of tax returns contain such errors? Perform the appropriate hypothesis test at the 5% level of significance.

Setting Up the Template

EXAMPLE

1. Type or import the labels from Column A of file **Chap07\Ex7-3.xls**. Select one cell in Column A. From the **Format** menu, choose **Column, Width**. Change the width to **30**. To format the headings, select Cells A1, A11, A17, and A22 and click **Bold** on the toolbar.

2. In Cells B3 and B4, enter the sample size and the number of successes in the sample, respectively. Using the above example, enter **500** in Cell B3 and **15** in Cell B4.

3. In Cell B5, enter the formula **=B4/B3** to compute the sample proportion.

4. In Cell B6, enter the desired hypothesized value (**0.02**). In Cell B8, enter the level of significance (**0.05**).

5. In Cell B7, enter the formula **=SQRT((B6*(1-B6))/B3)** to compute the standard deviation of the sample proportion.

6. In Cell B9, enter the formula **=(B5-B6)/B7** to compute the test statistic.

7. In Cell B12, enter the formula **=NORMSINV(1-B8/2)** to generate the upper critical value for the two-tailed hypothesis test.

8. In Cell B13, enter the formula **=NORMSINV(B8/2)** to generate the lower critical value for the two-tailed hypothesis test.

9. In Cell B14, enter the formula **=2*(1-NORMSDIST(ABS(B9)))** to compute the p-value for the two-tailed hypothesis test.

10. The decision rule involves a comparison of the p-value and the level of significance. In Cell B15, enter the formula **=IF(B14<B8,"Alternative","Null")**. If the value in Cell B14 (p-value) is less than the value in Cell B8 (alpha), then **Alternative** will be displayed in the cell; otherwise, **Null** will be displayed.

11. In Cell B18, enter the formula **=NORMSINV(1-B8)** to generate the critical value for the upper-tailed hypothesis test.

12. In Cell B19, enter the formula **=1-NORMSDIST(B9)** to compute the p-value for the upper-tailed hypothesis test.

13. The decision rule for an upper-tailed test follows the same logic as that of the two-tailed test. Enter the formula **=IF(B19<B8,"Alternative","Null")** in Cell B20.

14. In Cell B23, enter the formula **=NORMSINV(B8)** to generate the critical value for the lower-tailed hypothesis test.

15. In Cell B24, enter the formula **=NORMSDIST(B9)** to compute the *p*-value for the lower-tailed hypothesis test.

16. The decision rule for a lower-tailed test follows the same logic as that of the two-tailed test. Enter the formula **=IF(B24<B8,"Alternative","Null")** in Cell B25.

Discussing the Outcome

In all three tests, the null hypothesis is not rejected and **Null** is displayed in Cells 15, 20, and 25. The sample results support the claim that 2% of tax returns contain computational errors.

Now that the template has been set up, we can observe the effect of changes in the sample size, number of successes in the sample, and level of significance on the hypothesis test. For example, to determine the effect of a change in the number of successes on the test statistic, change the number of successes in Cell B4 from 15 to 20. Notice how the test statistic changes to 3.194, the two-tailed *p*-value becomes 0.0014, and the null hypothesis is rejected.

PROBLEMS

1. It is believed that 52% of Canadians drink coffee at least once a day. A consumer study asked 2000 Canadians about their habits and found that 49.5% said they drink coffee at least once day. At the 1% level of significance, determine if the study results are significantly different from the assumed proportion.

2. In the rapidly changing Canadian workplace, an increasing number of firms are offering child-care benefits to their employees. One union, however, claims that more than 90% of firms in the manufacturing sector still do not offer any child-care benefits to their workers. A random sample of 400 manufacturing firms found only 32 that actually did offer child-care benefits. Do the sample results support the union's claim? Test using a 10% level of significance.

PROBLEMS

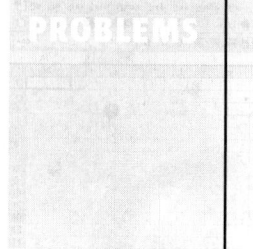

3. A human resources consultant to Unilever Canada claims that at most 40% of managers are dissatisfied with office politics. A random sample of 200 managers showed 90 who expressed dissatisfaction with office politics. Perform the appropriate hypothesis test at the 5% level of significance.

4. A vice-president of manufacturing claims that at most 3% of items produced on a certain assembly line are defective. A random sample of 100 items from the assembly line had 4 defectives. Perform the appropriate hypothesis test at the 5% level of significance.

5. The dean of the University of Alberta claims that no more than 10% of students drop out of university. She chooses a random sample of 500 students and finds that 56 of the 500 students dropped out. Do the sample results support the dean's claim? Test using a 5% level of significance.

CHAPTER 8
Hypothesis Testing: Two Populations

8.1 POOLED VARIANCE *t*-TEST AND CONFIDENCE INTERVAL

The pooled variance *t*-test for two population means is used when observations are taken from two separate populations. The random samples are therefore independent and do not have to be the same size. For example, one sample might consist of executives from Microsoft and the other of executives from General Motors. This method assumes that the populations are normally distributed and, since the test uses the sample standard deviations, the *t*-distribution will be required. It must also be assumed that the population variances, although unknown, are equal.

Hypothesis testing: difference between two population means (independent samples):

$$H_0: \mu_2 - \mu_1 = 0 \quad (\mu_2 = \mu_1)$$
$$H_1: \mu_2 - \mu_1 \neq 0 \quad (\mu_2 \neq \mu_1)$$

$$\text{Test statistic: } t^* = \frac{(\bar{x}_2 - \bar{x}_1) - (\mu_2 - \mu_1)}{\sigma(\bar{x}_2 - \bar{x}_1)}$$

If at least one of the sample sizes is less than 30, and if the population variances are unknown but assumed to be equal [use critical value: $\pm t(1 - \alpha/2; n_1 + n_2 - 2)$], then

$$\sigma(\bar{x}_2 - \bar{x}_1) = s_c \sqrt{\frac{1}{n_2} + \frac{1}{n_1}}$$

$$\text{where } s_c^2 = \frac{(n_1 - 1) s_1^2 + (n_2 - 1) s_2^2}{(n_1 - 1) + (n_2 - 1)} = \text{pooled estimate for } \sigma^2$$

Since the variances are assumed to be equal, they are also assumed to be common to the two populations. Therefore, a single estimate of the population variance is made by pooling the two sample variances together.

Confidence interval: difference between two population means (independent samples):

$$\mu_2 - \mu_1 = (\bar{x}_2 - \bar{x}_1) \pm t(1 - \alpha/2;\, n_1 + n_2 - 2)\, s_c \sqrt{\frac{1}{n_2} + \frac{1}{n_1}}$$

Excel has a **Data Analysis** tool for a pooled variance *t*-test but not for the confidence interval, so Excel functions and formulas can be used instead. The decision rule for the hypothesis test is based on the *p*-value approach. The decision rule is the following:

Do not reject H_0 if *p*-value $\geq \alpha$.
Reject H_0 if *p*-value $< \alpha$.

The steps for setting up the hypothesis test and confidence interval are shown in the following example.

EXAMPLE — REVENUE PER EMPLOYEE — Chap08\Ex8-1.xls

These days, top employers are judged not by the size of their payroll but by productivity as measured by revenue and/or profit per employee. The following table shows (in dollars) the revenue per employee for 12 companies with a head office in Montreal (Population #2) and 22 companies with a head office in Toronto (Population #1):

EXAMPLE

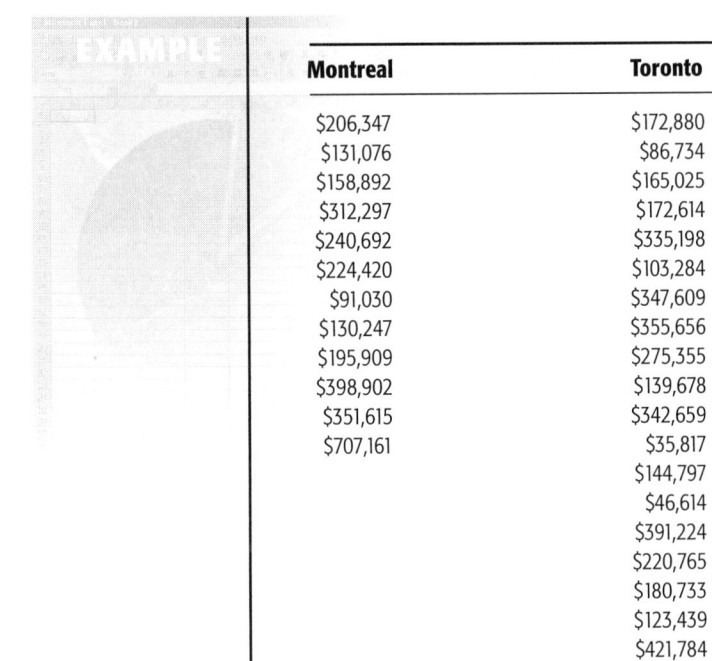

Montreal	Toronto
$206,347	$172,880
$131,076	$86,734
$158,892	$165,025
$312,297	$172,614
$240,692	$335,198
$224,420	$103,284
$91,030	$347,609
$130,247	$355,656
$195,909	$275,355
$398,902	$139,678
$351,615	$342,659
$707,161	$35,817
	$144,797
	$46,614
	$391,224
	$220,765
	$180,733
	$123,439
	$421,784
	$355,275
	$331,528
	$281,302

Source: Report on Business Magazine, July 1996.

Test the hypothesis that revenue per employee is the same for companies with head offices in Montreal and companies with head offices in Toronto at the 5% level of significance.

Setting Up the Template

1. Type or import the headings and data from Columns A and B of file **Chap08\Ex8-1.xls.** Select Cells A1:B1. From the **Format** menu, choose **Column, Width**. Change the width to **12**. To format the headings, select Cells A1:B1 and click **Bold** on the toolbar. Select Cells A3:B24 and, from the **Format** menu, choose **Cells**. On the **Number** tab, select the **Currency** category and click **OK**.

2. From the **Tools** menu, choose **Data Analysis**. From the **Data Analysis** tools list, choose **t-Test: Two-Sample Assuming Equal Variances** and click **OK**.

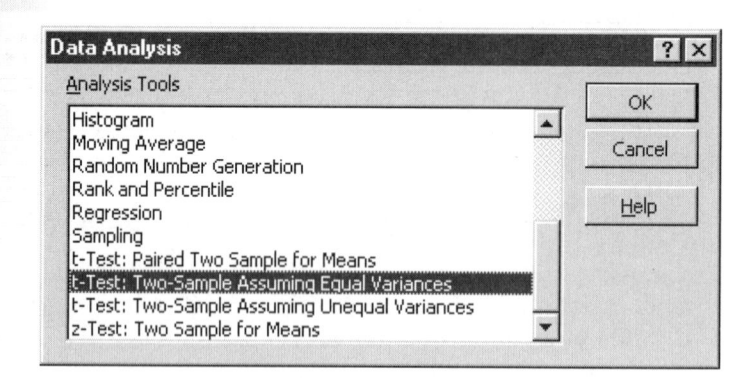

3. In the **t-Test: Two-Sample Assuming Equal Variances** dialog box, enter **A1:A14** for the **Variable 1 Range** and enter **B1:B24** for the **Variable 2 Range**. Since Excel assumes $H_0: \mu_1 - \mu_2 = 0$, we simply assign the values from Population #2 to the Variable 1 range and vice versa to be consistent with the above notation of $H_0: \mu_2 - \mu_1 = 0$. Enter **0** for **Hypothesized Mean Difference**, check the **Labels** check box, and enter the value for **Alpha**. Choose the **Output Range** button, then click in the box and enter **D1** for the **Output Range**. Click **OK**.

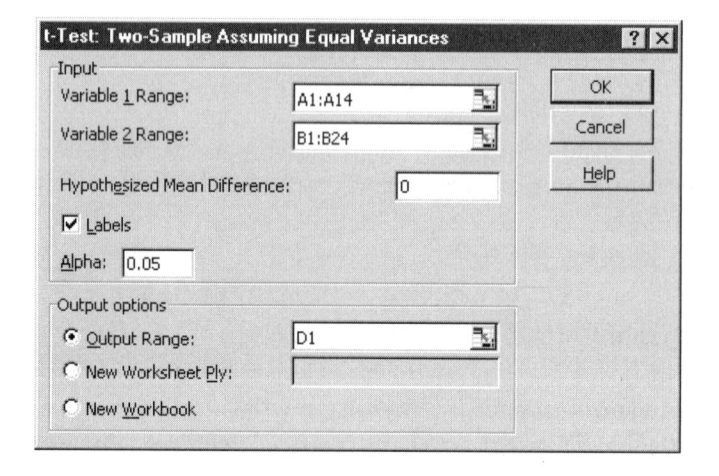

4. The results will be produced, but notice that some of the titles do not fit in the columns. Rather than individually formatting each column's width, select all the output cells and, from the **Format** menu, choose **Column, AutoFit Selection**. The output may be formatted further by changing the number of decimal places for numerical results or by italicizing headings, in the same way that you would format parts of any other spreadsheet.

5. To now produce the confidence interval, type or import the labels in Cells D17:D23. Select Cell D17 and click **Bold** on the toolbar.

6. In Cell E19, enter the formula **=E4-F4** to compute the mean difference.

7. In Cell E20, enter the formula **=E19/E10** to compute the standard deviation of the mean difference. This is done by dividing the mean difference by the test statistic.

8. In Cell E21, enter the formula **=TINV(0.05,E9)*E20** to compute the half-width of the confidence interval by multiplying the *t*-value by the standard deviation of the mean difference. This formula assumes 95% confidence. For a 99% confidence interval, enter 0.01 as the first parameter of the **TINV** function.

9. In Cell E22, enter the formula **=E19-E21** to compute the lower bound of the interval by subtracting the half-width from the mean difference.

10. In Cell E23, enter the formula **=E19+E21** to compute the upper bound of the interval by adding the half-width to the mean difference.

11. Select Cells E19:E23 and, from the **Format** menu, choose **Cells**. On the **Number** tab, select the **Currency** category and click **OK**.

Discussing the Outcome

The sample results reveal that there is not sufficient evidence to indicate a difference in the revenue per employee for companies with head offices in Montreal and companies with head offices in Toronto at the 5% level of significance. The *p*-value [P(T<=t) two-tail] of 0.4993 is greater than the level of significance of 0.05; also, the test statistic of 0.6834 falls between the critical values of ±2.037. The confidence interval estimate of $-\$66,843.21 \leq \mu_2 - \mu_1 \leq \$134,337.87$ is consistent with the conclusion of the hypothesis test, as 0 is included in the interval. There is a 95% probability that the population mean difference is between −$66,843.21 and $134,337.87.

Using Excel, we can observe the effect of changes in the raw data on the results. For example, increase all the values for Montreal by $100,000, then use the **t-Test: Two-Sample Assuming Equal Variances** tool on the revised data set. The sample results now show that there is sufficient evidence to indicate a difference in the revenue per employee for companies with head offices in Montreal and companies with head offices in Toronto at the 5% level of significance. The *p*-value [P(T<=t) two-tail] of 0.0108 is less than the level of significance of 0.05; also, the test statistic of 2.708 falls outside the critical values of ±2.037.

EXAMPLE

Note that whenever the raw data change, the **Data Analysis** tool must be used to get the updated results. The output of the **Data Analysis** tool is *not* automatically updated to reflect changes in the analysed data. On the other hand, the confidence interval constructed with formulas and functions is automatically updated to reflect changes in the analysed data. The confidence interval of $33,156.79 \leq \mu_2 - \mu_1 \leq \$234,337.87$ is consistent with the conclusion of the hypothesis test, as 0 is *not* included in the interval. There is a 95% probability that the population mean difference is between $33,156.79 and $234,337.87. Since the mean difference is based on (Montreal – Toronto), we can say that the revenue per employee for companies with head offices in Montreal is larger than that of companies with head offices in Toronto by between $33,156.79 and $234,337.87.

PROBLEMS

Chap08\P8-1#1.xls

1. The following table shows short-term debt as a percentage of total invested capital for firms in two industries. Assume that the ratios are normally distributed and that the populations are infinitely large. Find a 95% confidence interval for the difference between the means.

Automobile	Telecommunications
2.8	34.2
4.8	17.3
8.9	38.4
4.2	29.5
3.7	27.0
3.4	11.9
2.7	2.8
3.4	34.3
1.9	
3.8	
2.7	
2.6	
3.8	
3.6	
2.5	

Chap08\P8-1#2.xls

2. Random samples of the cost of second-hand statistics textbooks at two different bookstores have been selected with the following results (prices recorded to the nearest $0.10):

PROBLEMS

Studyhard Inc.
18.60 21.20 32.70 49.50 60.40 40.10 19.00 4.70 24.80 17.90 32.30

Goodgrades Co.
72.40 49.80 51.50 31.30 50.50 61.60 36.20 45.40 18.00 50.70 51.80
43.60 60.30 32.80 42.10

Can we conclude, at the 5% level of significance, that mean prices are different in the two bookstores?

3. A common measure for a firm's liquidity is the current ratio (the ratio of current assets to current liabilities). The table below presents the current ratios for 12 randomly selected companies in the food and lodging industry and for 15 randomly selected companies in the railroad industry.

Food and Lodging
1.5 0.8 1.2 1.6 1.4 2.6 1.6 1.1 0.7 1.1 1.0 1.3

Railroad
2.8 2.4 2.2 2.3 3.6 2.3 4.3 2.2 2.4 1.9 3.4 2.9
2.8 4.3 3.3

Can we conclude that the mean current ratio of the food and lodging industry (Population #1) is less than the mean current ratio of the railroad industry (Population #2)? Use $\alpha = 1\%$ and assume that the two populations are independent and normally distributed, and that the variances of current ratios for the two types of companies are equal.

4. Sales invoices from two salespeople, Elaine (Population #1) and George (Population #2), are to be compared. Five of Elaine's invoices and six of George's invoices are sampled with the following results (in dollars):

Elaine	George
1,300	2,200
800	3,000
300	700
1,500	1,800
800	600
	2,700

If the hypothesis to be tested is that the mean of Elaine's invoices is greater than the mean of George's invoices, what conclusion can be made at the 5% level of significance?

PROBLEMS

📄 Chap08\P8-1#5.xls

5. A Canadian insurance company has recorded the ages at which policy holders cashed in their retirement savings plans, with the following results:

Men	56	91	61	85	75	81	81	73	94
	74	58	72	68	52	77			
Women	75	67	72	55	93	52	81	75	86
	77	88	63						

If men are designated as Population #1 and women as Population #2, is there any difference in the mean ages at which men and women cash in their retirement savings plans, using a 95% confidence interval?

8.2 MATCHED SAMPLES *t*-TEST AND CONFIDENCE INTERVAL

The matched samples *t*-test for two population means is used when observations are taken from the same sample group twice. The samples are therefore not independent. One application is to record observations before and after some type of treatment. For instance, a group of students could be given a speed-reading test before and after taking a speed-reading course.

All observations are paired, and for each pair the difference is found $(d = x_2 - x_1)$.

Hypothesis testing: two population means (matched samples):

$$H_0: \mu_2 - \mu_1 = 0$$
$$H_1: \mu_2 - \mu_1 \neq 0$$

$$\text{Test statistic: } t^* = \frac{\bar{d} - \mu_d}{s_d/\sqrt{n}}$$

where $\bar{d}$ = the mean of the differences
μ_d = the hypothesized mean difference
s_d = the standard deviation of the differences
n = the number of pairs of observations

SECTION 8.2 • MATCHED SAMPLES *t*-TEST AND CONFIDENCE INTERVAL

Recall that $\bar{d} = \Sigma d / n$ and $s_d = \sqrt{\dfrac{\Sigma d^2 - (\Sigma d)^2/n}{n-1}}$.

Critical value: $\pm t(1 - \alpha/2; n - 1)$
Use $\pm z(1 - \alpha/2)$ if $n \geq 30$.

Decision rule: Conclude H_0 if $A_1 \leq t^* \leq A_2$.

Confidence interval: two population means (matched samples):

$$\mu_d = \bar{d} \pm t(1 - \alpha/2; n - 1)s_d/\sqrt{n}$$

Excel has a **Data Analysis** tool for a matched samples *t*-test but not for the confidence interval, so Excel functions and formulas can be used instead. The decision rule for the hypothesis test is based on the *p*-value approach. The decision rule is the following:

Do not reject H_0 if *p*-value $\geq \alpha$.
Reject H_0 if *p*-value $< \alpha$.

The steps for setting up the hypothesis test and confidence interval are shown in the following example.

EXAMPLE — VALUE OF CANADIAN EXPORTS   Chap08\Ex8-2.xls

A recent drop in the value of the Canadian dollar versus foreign currencies is expected to increase the value of Canadian exports. A comparison of the current year's shipments (Population #1) versus last year's shipments (Population #2) for each of seven Canadian exporters is given below (in $1,000s):

Exporter	Last Year	Current Year
1	4.25	4.84
2	5.92	5.01
3	2.63	2.90
4	3.91	4.26
5	4.88	5.17
6	3.16	3.15
7	4.22	3.99

114 ▶ CHAPTER 8 • HYPOTHESIS TESTING: TWO POPULATIONS

Assume that the exporters represent a random sample selected from all Canadian exporters. Do the data provide sufficient evidence to indicate a difference in the mean dollar value of exports between last year and the current year? Use a 5% level of significance.

Setting Up the Template

1. Type or import the data from Columns A, B, and C of file **Chap08\Ex8-2.xls**. Select Cells A1:C1. From the **Format** menu, choose **Column, Width**. Change the width to **12**. To format the headings, select Cells A1:C1 and click **Bold** on the toolbar. Select Cells A1:C8 and click **Center** on the toolbar.

2. From the **Tools** menu, choose **Data Analysis**. From the **Data Analysis** tools list, choose **t-Test: Paired Two Sample for Means** and click **OK**.

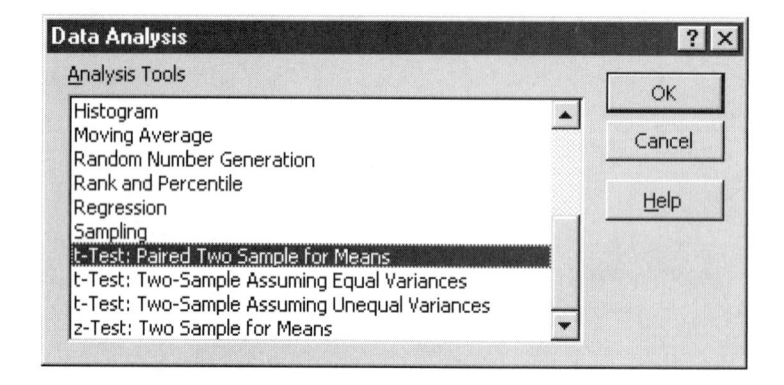

3. In the **t-Test: Paired Two Sample for Means** dialog box, enter **B1:B8** for the **Variable 1 Range** and enter **C1:C8** for the **Variable 2 Range**. Since Excel assumes $H_0: \mu_1 - \mu_2 = 0$, we simply assign the values from Population #2 to the Variable 1 range and vice versa to be consistent with the above notation of $H_0: \mu_2 - \mu_1 = 0$. Enter **0** for **Hypothesized Mean Difference**, check the **Labels** check box, and enter the value for **Alpha**. Choose the **Output Range** button, then click in the box and enter **E1** for the **Output Range**. Click **OK**.

EXAMPLE

[Screenshot of t-Test: Paired Two Sample for Means dialog box with Variable 1 Range B1:B8, Variable 2 Range C1:C8, Hypothesized Mean Difference 0, Labels checked, Alpha 0.05, Output Range E1]

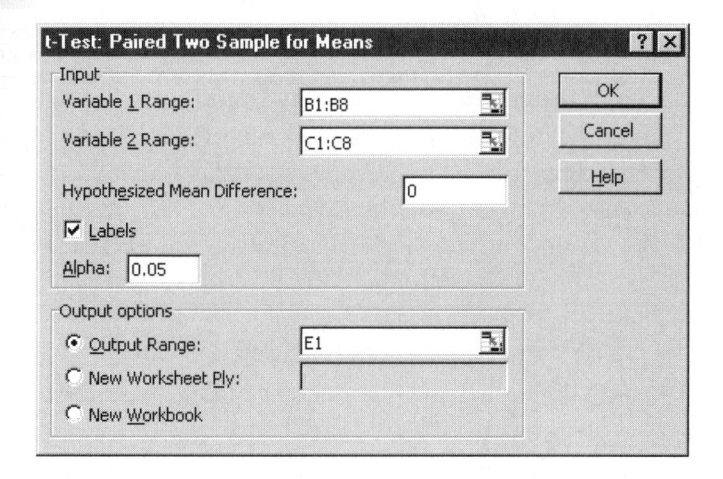

4. The results will be produced, but notice that some of the titles do not fit in the columns. Rather than individually formatting each column's width, select all the output cells and, from the **Format** menu, choose **Column, AutoFit Selection**. The output may be formatted further by changing the number of decimal places or italicizing headings, in the same way that you would format parts of any other spreadsheet.

5. To now produce the confidence interval, type or import the labels from Cells E17:E23. Select Cell E17 and click **Bold** on the toolbar.

6. In Cell F19, enter the formula **=F4-G4** to compute the mean difference.

7. In Cell F20, enter the formula **=F19/F10** to compute the standard deviation of the mean difference. Since $t^* = \bar{d}/s(\bar{d})$, then $s(\bar{d}) = \bar{d}/t^*$.

8. In Cell F21, enter the formula **=TINV(0.05,F9)*F20** to compute the half-width of the confidence interval by multiplying the t-value by the standard deviation of the mean difference. This formula assumes 95% confidence. For a 99% confidence interval, enter 0.01 as the first parameter of the **TINV** function.

9. In Cell F22, enter the formula **=F19-F21** to compute the lower bound of the interval by subtracting the half-width from the mean difference.

10. In Cell F23, enter the formula **=F19+F21** to compute the upper bound of the interval by adding the half-width to the mean difference.

Discussing the Outcome

The sample results reveal that there is not sufficient evidence to indicate a difference in the mean dollar value of exports between last year and the current year at the 5% level of significance. The *p*-value [P(T<=t) two-tail] of 0.7996 is greater than the level of significance of 0.05; also, the test statistic of −0.2654 falls between the critical values of ±2.447. The confidence interval of $-0.5111 \leq \mu_d \leq 0.4111$ is consistent with the conclusion of the hypothesis test, as 0 is included in the interval. There is a 95% probability that the population mean difference is between −0.5111 and 0.4111, in thousands of dollars.

Using Excel, we can observe the effect of changes in the raw data on the results. For example, increase all the values for the current year by 1 ($1,000), then use the **t-Test: Paired Two Sample for Means** tool on the revised data set. The sample results now show that there is sufficient evidence to indicate a difference in the mean dollar value of exports between last year and the current year at the 5% level of significance. The *p*-value [P(T<=t) two-tail] of 0.0014 is less than the level of significance of 0.05; also, the test statistic of −5.5724 falls outside the critical values of ±2.447.

Note that whenever the raw data change, the **Data Analysis** tool must be used to get the updated results. The output of the **Data Analysis** tool is *not* automatically updated to reflect changes in the data. On the other hand, the confidence interval constructed with formulas and functions is automatically updated to reflect changes in the analysed data. The confidence interval of $-1.5111 \leq \mu_d \leq -0.5889$ is consistent with the conclusion of the hypothesis test, as 0 is *not* included in the interval. There is a 95% probability that the population mean difference is between −1.5111 and −0.5889, in thousands of dollars. Since the mean difference is based on (last year − current year), we can say that the current year's exports are larger than last year's exports by between 0.5889 and 1.5111, in thousands of dollars.

PROBLEMS

Chap08\P8-2#1.xls

1. An investment analyst claims to have mastered the art of forecasting the price changes of a particular commodity. The following table gives the actual price changes (Population #1) and the changes forecast by the investment analyst (Population #2) in percent. Find a 95% confidence interval for the difference between the mean of the actual price changes and forecast changes.

PROBLEMS

Month	Actual Price Change	Forecast Price Change
1	7.5	14.2
2	−2.2	−18.6
3	8.7	7.3
4	−1.6	−5.0
5	9.9	1.8
6	6.3	−0.7
7	−4.4	−8.9
8	−0.1	6.9

Chap08\P8-2#2.xls

2. A tire manufacturer wishes to compare the wearing quality of two brands of tires: Goodyear (Population #1) and Michelin (Population #2). To make the comparison, a Goodyear tire and a Michelin tire were randomly assigned to the rear wheels of each of 10 cars (that is, each car had one Goodyear and one Michelin tire on the rear wheels). The cars were then driven for 10,000 miles and the amount of tread wear was recorded for each of the 20 tires with the following results:

Car	Goodyear Tire	Michelin Tire
1	10.5	10.3
2	9.9	9.8
3	12.2	11.6
4	9.6	9.4
5	8.7	8.5
6	11.8	11.2
7	10.6	10.7
8	12.3	12.4
9	9.0	9.3
10	11.1	11.0

Test the hypothesis that Michelin tires have less tread wear on the average than Goodyear tires, at the 1% level of significance.

3. A random sample of male students from the University of New Brunswick is selected and reaction times to a certain stimulus are measured before and after drinking two bottles of beer. The results are as follows (in seconds):

Before	**(Population #1)**	4.1	3.7	3.8	4.3	3.1
After	**(Population #2)**	4.5	3.8	4.1	4.4	3.3

Test the hypothesis that the mean reaction time is slower after drinking the beer, at a 1% level of significance and assuming that reaction times are normally distributed.

PROBLEMS

Chap08\P8-2#4.xls

4. Research on the effectiveness of a thyroid medication has generated the following efficiency results for a random sample of 12 patients who were put on the medication for a month:

Patient	1	2	3	4	5	6	7	8	9	10	11	12
Before (X)	135	141	123	121	132	130	125	121	137	122	118	105
After (Y)	127	139	115	110	124	128	119	117	122	116	103	101

It is assumed that the population of differences $(Y - X)$ is approximately normal. What is the 95% confidence interval estimate for the mean difference?

Chap08\P8-2#5.xls

5. A study was conducted in Toronto to compare the mean supermarket prices of two leading brands of soda. Ten supermarkets were selected, and the price of a six-pack of canned soda was recorded for each brand. Do the data provide sufficient evidence to indicate a difference in the mean prices of a six-pack for the two brands of soda? Test using $\alpha = 5\%$.

SUPERMARKET	PRICE Brand 1	Brand 2
1	$2.27	$2.31
2	2.42	2.41
3	2.39	2.45
4	2.17	2.25
5	2.25	2.29
6	2.22	2.23
7	2.30	2.40
8	2.35	2.44
9	2.28	2.37
10	2.54	2.50

8.3 z-TEST FOR DIFFERENCES IN TWO PROPORTIONS

Hypothesis testing: difference in two proportions (independent samples):

$$H_0: p_2 - p_1 = 0 \quad (p_2 = p_1)$$
$$H_1: p_2 - p_1 \neq 0 \quad (p_2 \neq p_1)$$

Test statistic: $z^* = \dfrac{(\bar{p}_2 - \bar{p}_1) - (p_2 - p_1)}{s(\bar{p}_2 - \bar{p}_1)}$

Critical value: $\pm z(1 - \alpha/2)$
For upper-tailed tests: $z(1 - \alpha)$
For lower-tailed tests: $-z(1 - \alpha)$

$$s(\bar{p}_2 - \bar{p}_1) = \sqrt{\bar{p}'(1 - \bar{p}')\left(\dfrac{1}{n_1} + \dfrac{1}{n_2}\right)}$$

where pooled estimate for $p = \bar{p}' = \dfrac{n_1 \bar{p}_1 + n_2 \bar{p}_2}{n_1 + n_2} = \dfrac{x_1 + x_2}{n_1 + n_2}$

x_1, x_2 = the number of items in each sample with the given characteristic
$\bar{p}_1, \bar{p}_2$ = the sample proportions

Excel does not have a **Data Analysis** tool to test a hypothesis for the difference in two population proportions, so Excel functions and formulas can be used instead. The decision rule for the test is based on the *p*-value approach. For each of the three types of hypothesis tests (upper-tailed, lower-tailed, and two-tailed), the decision rule is the following:

Do not reject H_0 if *p*-value $\geq \alpha$.
Reject H_0 if *p*-value $< \alpha$.

To test the hypothesis for the difference in two population proportions, we must have the sample size and number of successes for each sample, the level of significance, and the hypothesized difference in the null hypothesis. These can be entered as values or, if the sample size and number of successes must be obtained from raw data, the **PivotTable Wizard** could be used and the results copied into the appropriate cells. The steps for setting up the hypothesis test are shown in the following example.

EXAMPLE: SUPPORT FOR USER FEES

Chap08\Ex8-3.xls

A random sample of 160 doctors in Quebec (Population #1) was found to have 88 who oppose user fees for Medicare services. An independent random sample of 240 doctors in Ontario (Population #2) had 108 who oppose user fees. Is there sufficient evidence to reject the claim that equal proportions of doctors in the two provinces oppose user fees? Use a 10% level of significance.

Setting Up the Template

1. Type or import the labels from Column A of file **Chap08\Ex8-3.xls**. Select one cell in Column A. From the **Format** menu, choose **Column, Width**. Change the width to **30**. Select one cell in Column B. From the **Format** menu, choose **Column, Width**. Change the width to **12**. To format the headings, select Cells A1, A3, A8, A19, A25, and A30 and click **Bold** on the toolbar.

2. In Cells B4, B5, B9, and B10, enter the number of successes and sample size for Sample #1 and Sample #2, respectively. Using our example, enter **88** and **160** in Cells B4:B5, and **108** and **240** in Cells B9:B10.

3. In Cell B6, enter the formula **=B4/B5** to compute the sample proportion for Sample #1. In Cell B11, enter the formula **=B9/B10** to compute the sample proportion for Sample #2.

4. In Cell B13, enter the formula **=(B4+B9)/(B5+B10)** to compute the pooled estimate for p.

5. In Cell B14, enter the formula **=B11-B6** to compute the difference in the two sample proportions ($\bar{p}_2 - \bar{p}_1$).

6. In Cells B15 and B16, enter the desired hypothesized difference (**0**) and level of significance (**0.1**).

7. Enter the formula **=(B14-B15)/SQRT(B13*(1-B13)*(1/B5+1/B10))** to compute the test statistic in Cell B17.

8. In Cell B20, enter the formula **=NORMSINV(B16/2)** to generate the lower critical value for the two-tailed hypothesis test.

9. In Cell B21, enter the formula **=NORMSINV(1-B16/2)** to generate the upper critical value for the two-tailed hypothesis test.

10. In Cell B22, enter the formula **=2*(1-NORMSDIST(ABS(B17)))** to compute the p-value for the two-tailed hypothesis test.

11. The decision rule involves a comparison of the p-value and the level of significance. In Cell B23, enter the formula **=IF(B22<B16,"Alternative","Null")**. If the value in Cell B22 (p-value) is less than the value in Cell B16 (alpha), then **Alternative** will be displayed in the cell; otherwise, **Null** will be displayed.

12. In Cell B26, enter the formula **=NORMSINV(B16)** to generate the critical value of the lower-tailed hypothesis test.

13. In Cell B27, enter the formula **=NORMSDIST(B17)** to compute the p-value for the lower-tailed hypothesis test.

14. The decision rule for a lower-tailed test follows the same logic as that of the two-tailed test. Enter the formula **=IF(B27<B16,"Alternative","Null")** in Cell B28.

15. In Cell B31, enter the formula **=NORMSINV(1-B16)** to generate the critical value of the upper-tailed hypothesis test.

16. In Cell B32, enter the formula **=1-NORMSDIST(B17)** to compute the p-value for the upper-tailed hypothesis test.

17. The decision rule for an upper-tailed test follows the same logic as that of the two-tailed test. Enter the formula **=IF(B32<B16,"Alternative","Null")** in Cell B33.

Discussing the Outcome

The sample results indicate that there is sufficient evidence to reject the claim that equal proportions (two-tailed test) of doctors in the two provinces oppose user fees at the 10% level of significance, since 0.049998 < 0.1.

Now that the template has been set up, we can observe the effect of changes in the number of successes or sample size for a particular sample and in the level of significance on the hypothesis test. For example, to determine the effect of a

EXAMPLE

change in the number of doctors in Ontario who oppose user fees on the conclusion of the test, change the value in Cell B9 from 108 to 120. The sample proportion for Ontario changes to 0.5, which in turn changes the pooled estimate to 0.52, the difference in the sample proportions to −0.05, the test statistic to −0.98, and the p-value for the two-tailed hypothesis test to 0.3268. The null hypothesis is not rejected in this situation, as the p-value is greater than 0.1; also, the test statistic falls between the critical values of ± 1.645. In this case, there is not sufficient evidence to reject the claim that equal proportions of doctors in the two provinces oppose user fees at the 10% level of significance.

PROBLEMS

1. Research is undertaken to determine the effectiveness of a new medication. Three hundred patients who are diagnosed as having a certain disease are randomly assigned to two groups—the research group (Population #1), consisting of 100 patients, and the control group (Population #2), which has the remaining 200 patients. Patients in the research group are given the new medication, while patients in the control group are not given the medication. In all other respects, the two groups are treated identically. After one year, it is found that 75 patients in the research group have been cured and 120 patients in the control group have been cured. Test the hypothesis that the medication is effective in curing the disease at the 1% level of significance.

2. After three years of deficit cutting by Alberta's provincial government, public concerns about health care and hospital closures seem to be on the rise. A random sample of 833 Albertans polled in February/March 1995 (Population #1) had 26% who replied "health care and hospital closures" when asked what was the most important issue facing Alberta. A random sample of 800 Albertans polled in May 1996 (Population #2) had 50% who replied "health care and hospital closures." Test the hypothesis that a higher proportion of Albertans were of the opinion that health care and hospital closures was the most important issue in May 1996 than in February/March 1995. Use $\alpha = 5\%$. (*Source: Calgary Herald*, Saturday, June 15, 1996.)

3. In a random sample of 200 PC owners (Population #1), 30 are university students. In a random sample 160 Mac owners (Population #2), 15 are university students. Perform a hypothesis test to determine whether the proportion of university student owners is larger for the PC than for the Mac. Use a 5% level of significance.

PROBLEMS

4. A random sample of 150 accountants (Population #1) had 30% who expressed dissatisfaction with their choice of career. A random sample of 200 lawyers (Population #2) had 40% who expressed dissatisfaction. Test the hypothesis that a higher proportion of lawyers than accountants are dissatisfied with their choice of career.

5. A company produces pesticides. A new product called Deadfly (Population #1) is to be compared with the product currently considered the market leader, Lites-Out (Population #2). Two rooms of equal size are sprayed with the same amount of spray, one room with Deadfly and the other with Lites-Out. Three hundred mosquitoes are released into one room and four hundred into the other. After 30 minutes, the dead insects are counted. The results are as follows:

	Deadfly	**Lites-Out**
Number of insects released	300	400
Number of dead insects	195	275

Test whether the new product is more effective than the market leader. Use $\alpha = 5\%$.

8.4 F-TEST FOR TWO POPULATION VARIANCES

Here we will discuss a hypothesis test to compare the variances of two independent samples. The sampling distribution of the test statistic is the F-distribution, which has a pair of degrees of freedom, one for the numerator and one for the denominator. The test is commonly used to see if the two population variances can be assumed to be equal or unequal.

$$H_0: \sigma_2^2 / \sigma_1^2 = 1 \quad (\sigma_2^2 = \sigma_1^2)$$
$$H_1: \sigma_2^2 / \sigma_1^2 \neq 1 \quad (\sigma_2^2 \neq \sigma_1^2)$$

Test statistic: $F^* = s_2^2 / s_1^2$
Critical value: $A_1 = F(\alpha/2; n_2 - 1; n_1 - 1)$
$A_2 = F(1 - \alpha/2; n_2 - 1; n_1 - 1)$
For lower-tailed tests: $A = F(\alpha; n_2 - 1; n_1 - 1)$
For upper-tailed tests: $A = F(1 - \alpha; n_2 - 1; n_1 - 1)$

Decision rule: Conclude H_0 if $A_1 \leq F^* \leq A_2$

Excel has a **Data Analysis** tool to perform an *F*-test for two population variances if the raw data is known. The decision rule for the hypothesis test is based on the *p*-value approach. The decision rule is the following:

$$\text{Do not reject } H_0 \text{ if } p\text{-value} \geq \alpha.$$
$$\text{Reject } H_0 \text{ if } p\text{-value} < \alpha.$$

The steps for setting up the hypothesis test are shown in the following example.

EXAMPLE — STOCK DIVIDEND YIELDS — Chap08\Ex8-4.xls

A researcher wishes to determine whether there is any difference between the variance of dividend yields of stocks traded on the Montreal Stock Exchange (Population #2) and those traded on the Toronto Stock Exchange (Population #1). A random sample of 20 companies from the MSE and a random sample of 24 companies from the TSE are selected. The results are shown in the following table:

MSE	TSE
3.3	1.3
2.5	5.2
5.4	4.0
2.0	0.9
3.1	3.3
3.4	2.8
3.2	3.6
3.9	1.4
1.7	2.5
2.8	0.8
3.5	2.7
6.3	1.2
5.5	2.0
2.9	1.9
3.0	2.6
5.1	4.6
1.1	2.7
2.4	1.3
3.8	3.5
2.6	2.2
	2.2
	1.6
	2.7
	2.4

SECTION 8.4 • F-TEST FOR TWO POPULATION VARIANCES

EXAMPLE

Test the hypothesis that the population variances are equal at the 5% level of significance.

Setting Up the Template

1. Type or import the data from Columns A and B of file **Chap08\Ex8-4.xls**. To format, select Cells A1:B1 and click **Bold** on the toolbar. Select Cells A1:B25 and click **Center** on the toolbar. Also using the toolbar, adjust the number of decimal places if desired, so that final zeros appear.

2. From the **Tools** menu, choose **Data Analysis**. From the **Data Analysis** tools list, choose **F-Test Two-Sample for Variances** and click **OK**.

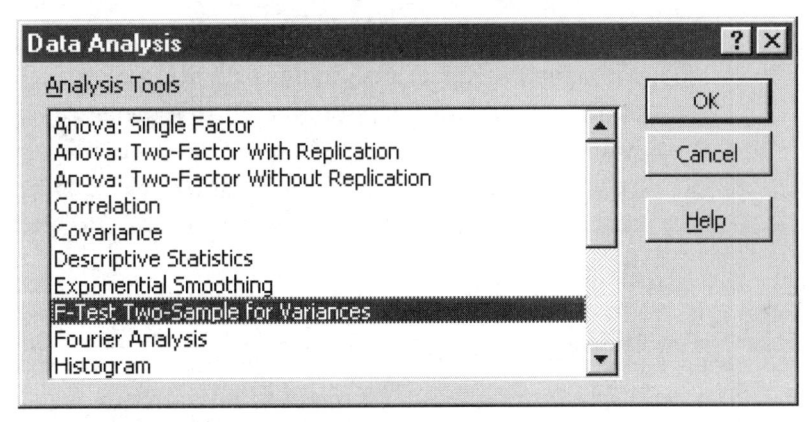

3. In the **F-Test Two-Sample for Variances** dialog box, enter **A1:A21** for the **Variable 1 Range** and **B1:B25** for the **Variable 2 Range**. Since Excel assumes H_0: $\sigma_1^2 / \sigma_2^2 = 1$ ($\sigma_1^2 = \sigma_2^2$), we simply assign the values from Population #2 to the Variable 1 range and vice versa to be consistent with the above notation of H_0: $\sigma_2^2 / \sigma_1^2 = 1$. Check the **Labels** check box, and enter the value for **Alpha**. Choose the **Output Range** button, then click in the box and enter **D1** for the **Output Range**. Click **OK**.

EXAMPLE

[F-Test Two-Sample for Variances dialog box: Variable 1 Range: A1:A21; Variable 2 Range: B1:B25; Labels checked; Alpha: 0.05; Output Range: D1]

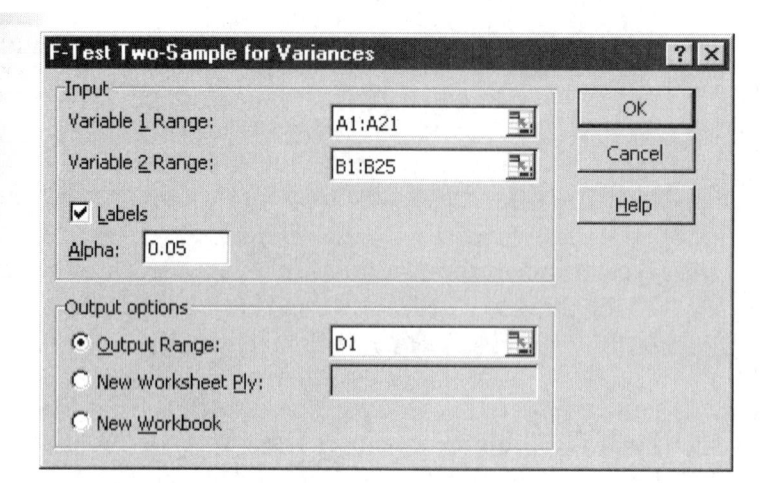

4. The results will be produced, but notice that some of the titles do not fit in the columns. Rather than individually formatting each column's width, select all the output cells and, from the **Format** menu, choose **Column, AutoFit Selection**. The output may be formatted further by changing the number of decimal places of numerical results or by italicizing headings, in the same way that you would format parts of any other spreadsheet.

Discussing the Outcome

The sample results reveal that there is not sufficient evidence to indicate a difference in the variances of dividend yields of stocks traded on the MSE and those traded on the TSE at the 5% level of significance. Since only the *p*-value for a one-tail test [P(F<=f) one-tail] is provided in the output and we are testing a two-sided alternative, we have to multiply our one-tail *p*-value by 2. We cannot reject the null hypothesis that the variances are equal, since the two-sided *p*-value of 0.47466 (0.23733 × 2) is greater than the level of significance of 0.05.

Using Excel, we can observe the effect of changes in the raw data on the results. For example, multiply all the values for the MSE by a factor of 1.5, then use the **F-Test Two-Sample for Variances** tool on the revised data set. The sample results now indicate that there is sufficient evidence to indicate a difference in the variances of dividend yields of stocks traded on the MSE and those traded on the TSE at the 5% level of significance. The *p*-value [P(F<=f) one-tail] of 0.00584 multiplied by two is 0.01168, which is less than the level of significance of 0.05. Note that whenever the raw data change, the **DataAnalysis** tool must be used to get the updated results. The output of the **Data Analysis** tool is not automatically updated to reflect changes in the data.

PROBLEMS

Chap08\P8-4#1.xls

1. A production manager wants to test the hypothesis that the variance of the cost of producing polyfibres is the same for two different production processes. Random samples of production costs for several production runs using the two different processes are as follows (in dollars):

Process I	20	15	20	23	24	21
Process II	27	19	41	30	16	

Test the hypothesis that the population variances are equal at the 2% level of significance.

Chap08\P8-4#2.xls

2. Sales invoices from two sales people, Elaine (Population #1) and George (Population #2), are to be compared. Five of Elaine's invoices and six of George's invoices are sampled with the following results (in dollars):

Elaine	George
1,250	2,500
750	3,160
220	670
1,680	1,770
910	690
	2,830

If the hypothesis to be tested is that the variance of Elaine's invoices is less than the variance of George's invoices, what conclusion can be made at the 5% level of significance?

Chap08\P8-4#3.xls

3. Ten runners from a Canadian high school (Population #1) and ten runners from an American high school (Population #2) competed at a track meet. In the one-mile race, the runners' times were the following (in seconds):

Team A	248	245	252	251	262	260	266	275	274
Team B	240	251	253	258	277	276	278	272	289

Assume that these times represent independent random samples from the two populations. Test the hypothesis that the population variances are equal at the 10% level of significance.

CHAPTER 9
Chi-Square Tests

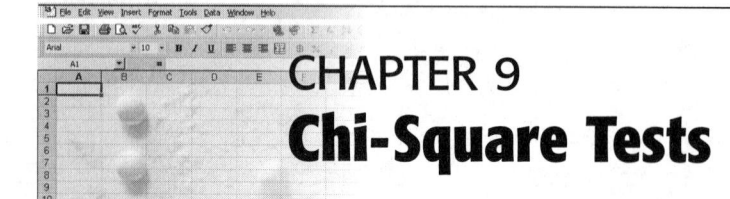

9.1 CHI-SQUARE TEST FOR NORMALITY

For the chi-square test, the null and alternative hypotheses are

H_0: normal distribution
H_1: not a normal distribution

Assuming that the null hypothesis is true, calculate the expected frequencies using the normal distribution:

$$E_i = P_i \times n$$

Test statistic = $\sum[(\text{observed} - \text{expected})^2 / \text{expected}] = (O_i - E_i)^2 / E_i$

Critical value: $\chi^2(1 - \alpha;\ df)$
where df = (# of classes) − (# of parameters estimated) − 1

Decision rule: Conclude H_0 if test statistic ≤ critical value.
Conclude H_1 if test statistic > critical value.

Excel does not have a **Data Analysis** tool to perform a chi-square test for normality, so Excel functions and formulas can be used instead. The decision rule for the test is based on the *p*-value approach. The decision rule is the following:

Do not reject H_0 if *p*-value ≥ α.
Reject H_0 if *p*-value < α.

To test for normality, we must have the observed frequencies for all class intervals as well as the level of significance. These can be entered as values or, if the observed frequencies must be obtained from raw data, the **PivotTable Wizard** could be used and the results copied into the appropriate cells. The steps for setting up the hypothesis test are shown in the following example.

EXAMPLE

APTITUDE TEST SCORES

Chap09\Ex9-1.xls

A personnel manager at Bell Canada claims that scores on an aptitude test are normally distributed. A sample of 1000 candidates had a mean score of 72 and a standard deviation of 10. The scores of the 1000 candidates were as follows:

Scores	O_i
< 50	25
50 to < 60	120
60 to < 70	310
70 to < 80	370
80 to < 90	150
90 and up	25
Total	1000

Test the claim at $\alpha = 5\%$.

Setting Up the Template

1. Type or import the labels from Cells A1:A3, A5:F6, and A16:A21 of file **Chap09\Ex9-1.xls**. Select the labels and click **Bold** and **Center** on the toolbar. Select cells A1:F1. From the **Format** menu, choose **Column, Width**. Change the width to **15**.

2. In Cells B1:B3, enter the sample size, mean, and standard deviation, respectively. Using our example, enter **1000, 72,** and **10**, respectively.

3. In Cells A7:A12, enter the lower limits of the class intervals. In Cells B7:B12, enter the upper limits of the class intervals.

4. In Cells C7:C12, type or import the observed frequencies.

5. To compute the normal probabilities for the class intervals, enter the function **=NORMDIST(B7,$B$2,$B$3,TRUE)** in Cell D7. B7 is the value for which we want the distribution, $B$2 is the mean, $B$3 is the standard

deviation, and TRUE returns the cumulative probability. Enter the function **=NORMDIST(B8,$B$2,$B$3,TRUE)-NORMDIST(A8,$B$2,$B$3,TRUE)** in Cell D8. Copy Cell D8 and paste to Cells D9:D11. Finally, enter the function **=1-NORMDIST(A12,$B$2,$B$3,TRUE)** in Cell D12.

6. To compute the expected frequencies, enter the formula **=D7*$B$1** in Cell E7. Copy Cell E7 and paste to Cells E8:E12.

7. In Cell F7, enter the formula **=(C7-E7)^2/E7**. Copy Cell F7 and paste to Cells F8:F12.

8. In Cell E14, enter the label **Test Statistic**. To compute the test statistic, enter the function **=SUM(F7:F12)** in Cell F14.

9. In Cell B16, enter the degrees of freedom (**3**, which comes from 6 – 2 – 1).

10. In Cell B17, enter the function **=CHIDIST(F14,B16)** to compute the p-value for test for normality. F14 refers to the test statistic and B16 refers to the degrees of freedom.

11. In Cell B18, enter the level of significance (**0.05**).

12. In Cell B19, enter the formula **=F14** to display the test statistic once again.

13. In Cell B20, enter the function **=CHIINV(B18,B16)** to compute the critical value with α = 5% and 3 degrees of freedom.

14. The decision rule involves a comparison of the p-value and the level of significance. In Cell B21, enter the formula **=IF(B17<B18,"Alternative","Null")**. If the value in Cell B17 (p-value) is less than the value in Cell B18 (alpha), then **Alternative** will be displayed in the cell; otherwise, **Null** will be displayed.

Discussing the Outcome

The sample results indicate that there is sufficient evidence to reject the manager's claim that the scores on an aptitude test are normally distributed at the 5% level of significance. The p-value of 0.0002 is less than the level of significance of 0.05; also, the test statistic of 19.59 is greater than the critical value of 7.81.

Now that the template has been set up, we can observe the effect of changes in the observed frequencies and in the level of significance on the hypothesis test. For example, to determine the effect of a change in the observed frequencies, change the observed frequencies in Cells C7:C12 to 15, 95, 315, 380, 165, and 30. Notice how the p-value changes from 0.0002 to 0.4176 and the test statistic changes from 19.59 to 2.84. The null hypothesis is not rejected in this situation, as the p-value is greater than the level of significance of 0.05; also, the test statistic falls below the critical value of 7.81. In this case, we would conclude that there is sufficient evidence to support the manager's claim that the scores on an aptitude test are normally distributed at the 5% level of significance.

The same approach could be used in other "goodness of fit" tests for the Poisson, binomial, or other types of distribution. In these situations, Excel's **POISSON** or **BINOMDIST** functions would be used in place of the **NORMDIST** function in computing the probabilities for the class intervals.

PROBLEMS

1. A production process is intended to have a daily output that is normally distributed. In order to test the hypothesis that output is normally distributed, a random sample of 105 days was selected and the following results were obtained:

Output (units)	Observed Frequency
Less than 300	3
300 to < 350	6
350 to < 400	8
400 to < 450	19
450 to < 500	27
500 to < 550	21
550 to < 600	12
600 and over	9

   The population mean and standard deviation are unknown, but assume a sample mean of 500 and a sample standard deviation of 100 units. Complete the appropriate hypothesis test at the 10% level of significance.

2. The duration of 100 randomly sampled long-distance telephone calls using Sprint Canada had the following frequency distribution:

Time (minutes)	Observed Frequency
Less than 2	6
2 to < 3	16
3 to < 4	32
4 to < 5	23
5 to < 6	16
6 or more	7

At the 5% level of significance, test the hypothesis that the duration of long-distance telephone calls has a normal distribution.

3. The associate dean of the University of British Columbia believes that the number of students dropping a class is Poisson distributed. Five hundred classes, all containing the same number of students, were randomly selected. The number of withdrawals from the classes was recorded. Assume the mean of the sample below is 4. What conclusion can be drawn about whether or not these data come from a Poisson distribution? Use a significance level of 5% to justify your conclusion.

Drops	Classes
0	0
1	20
2	60
3	100
4	180
5	40
6	60
7	40
>7	0
Total	500

9.2 CONTINGENCY TABLE TEST FOR INDEPENDENCE

In a contingency table, the data are classified according to two criteria. The objective is to see if the two criteria are independent of each other:

H_0: The two criteria are independent of each other.
H_1: The two criteria are *not* independent of each other.

The following notation is used for contingency tables:

RT_i = row total for row i
CT_j = column total for column j
NR = number of rows

NC = number of columns
O_{ij} = observed frequency for (i, j) cell
E_{ij} = expected frequency for (i, j) cell
n = sample size

Assuming that the null hypothesis is true, calculate the expected frequencies:

$$E_{ij} = P(\text{value found in } i\text{-}j \text{ cell}) \times n = P(\text{joint}) \times n$$

Recall that the following property of statistical independence should be true under H_0: $P(A \cap B) = P(A) \times P(B)$. From this:

$$E_{ij} = P(\text{row}) \times P(\text{column}) \times n = (RT_i/n) \times (CT_j/n) \times n = \frac{RT_i \times CT_j}{n}$$

$$\text{Test statistic} = \sum \frac{(O_{ij} - E_{ij})^2}{E_{ij}}$$

Critical value: $\chi^2(1 - \alpha; \text{df})$, where df = $(NR - 1)(NC - 1)$

Decision rule: Conclude H_0 if test statistic $\leq$ critical value.
Conclude H_1 if test statistic $>$ critical value.

Excel does not have a **Data Analysis** tool to perform a contingency table for independence, so Excel functions and formulas can be used instead. The decision rule for the test is based on the *p*-value approach. The decision rule is the following:

Do not reject H_0 if *p*-value $\geq \alpha$.
Reject H_0 if *p*-value $< \alpha$.

To test for independence, we must have the observed frequencies for all cells of the contingency table as well as the level of significance. These can be entered as values or, if the observed frequencies must be obtained from raw data, the **PivotTable Wizard** could be used and the results copied into the appropriate cells. The steps for setting up the test of hypothesis are shown in the following example.

EXAMPLE

UNDERGRADUATE SMOKERS

Chap09\Ex9-2.xls

A teacher at McGill University in Montreal claims that the amount of smoking done by undergraduate students depends on the year of study. A random sample of 200 students registered in a three-year bachelor's program were surveyed with the following results:

	YEAR		
SMOKING	**1**	**2**	**3**
Heavy	24	27	32
Light	14	47	17
Nonsmoker	7	22	10

Test the teacher's claim at the 5% level of significance.

Setting Up the Template

1. Type or import the labels from Column A of file **Chap09\Ex9-2.xls**. Select one cell in Column A. From the **Format** menu, choose **Column, Width**. Change the width to **25**. Select Cells B1:E1. From the **Format** menu, choose **Column, Width**. Change the width to **10**. Enter the headings **Year 1, Year 2, Year 3**, and **Total** in Cells B1:E1 and also in B7:E7. Select Cells A1:A20, B1:E1, and B7:E7 and click **Bold** on the toolbar. Select Cells B1:E11 and click **Center** on the toolbar.

2. In Cells B2:D4, type the observed frequencies in the table above or import them from file **Chap09\Ex9-2.xls**.

3. In Cell E2, enter the function **=SUM(B2:D2)**. Copy Cell E2 and paste to Cells E3:E4 to obtain the row totals.

4. In Cell B5, enter the function **=SUM(B2:B4)**. Copy Cell B5 and paste to Cells C5:E5 to obtain the column totals.

5. To compute the expected frequencies, enter the formula **=E2*$B$5/$E$5** in Cell B8, enter the formula **=E2*$C$5/$E$5** in Cell C8, and enter the formula **=E2*$D$5/$E$5** in Cell D8. Copy Cells B8:D8 and paste to Cells B9:D10. These calculations can be extended to contingency tables of different sizes.

EXAMPLE

6. In Cell E8, enter the function **=SUM(B8:D8)**. Copy Cell E8 and paste to Cells E9:E10 to obtain the row totals.

7. In Cell B11, enter the function **=SUM(B8:B10)**. Copy Cell B11 and paste to Cells C11:E11 to obtain the column totals.

8. In Cell B13, enter the function **=CHITEST(B2:D4,B8:D10)** to compute the p-value for the test for independence. Cells B2:D4 refer to the observed frequencies and Cells B8:D10 refer to the expected frequencies.

9. In Cells B14 and B15, enter the number of rows and the number of columns, respectively. Using our example, enter **3** in both cells.

10. In Cell B16, enter the formula **=(B14-1)*(B15-1)** to compute the degrees of freedom.

11. In Cell B17, enter the level of significance (**0.05**).

12. In Cell B18, enter the function **=CHIINV(B13,B16)** to compute the test statistic.

13. In Cell B19, enter the function **=CHIINV(B17,B16)** to compute the critical value with α = 5% and 4 degrees of freedom.

14. The decision rule involves a comparison of the p-value and the level of significance. In Cell B20, enter the formula **=IF(B13<B17,"Alternative","Null")**. If the value in Cell B13 (p-value) is less than the value in Cell B17 (alpha), then **Alternative** will be displayed in the cell; otherwise, **Null** will be displayed.

Discussing the Outcome

The sample results indicate that there is sufficient evidence to support the claim that the amount of smoking done by undergraduate students depends on their year of study at the 5% level of significance. The p-value of 0.0079 is less than the level of significance of 0.05; also, the test statistic of 13.82 is greater than the critical value of 9.49.

Now that the template has been set up, we can observe the effect of changes in the observed frequencies and in the level of significance on the hypothesis test. For example, to determine the effect of a change in the number of heavy smokers in Year 3 from 32 to 22 and in the number of light smokers in Year 3 from 17 to 27, change the value in Cell D2 from 32 to 22 and the value in Cell D3 from 17 to 27.

EXAMPLE

Notice how the *p*-value changes from 0.0079 to 0.0675 and the test statistic changes from 13.82 to 8.76. The null hypothesis is not rejected in this situation, as the *p*-value is greater than the level of significance of 0.05; also, the test statistic falls below the critical value of 9.49. We would conclude that there is sufficient evidence to reject the claim that the amount of smoking done by undergraduate students depends on year of study at the 5% level of significance.

PROBLEMS

1. A random sample of 406 students at Bishop's University were surveyed in order to investigate attitudes to a new university policy regarding final examinations. The results are seen in the following table:

	FACULTY		
OPINION	**Arts**	**Science**	**Management**
Favour	54	47	91
Oppose	36	38	82
Undecided	25	6	27

Perform a test of hypothesis to determine whether opinion is independent of faculty at the 5% level of significance.

2. A financial consultant is interested in whether the differences in capital structure are related to firm size in a certain Canadian industry. The consultant surveys a group of firms with assets of different amounts and divides the firms into three groups by asset size. Each firm is classified according to whether its total debt is greater than or less than stockholders' equity. The results of the survey are as follows:

	FIRM ASSET SIZE (in thousands)			
	< $500	**$500 to $2,000**	**> $2,000**	**Total**
Debt < equity	9	12	5	**26**
Debt > equity	7	10	21	**38**
Total	**16**	**22**	**26**	**64**

Test the consultant's hypothesis at the 10% level of significance.

SECTION 9.2 • CONTINGENCY TABLE TEST FOR INDEPENDENCE

PROBLEMS

Chap09\P9-2#3.xls

3. A local market research magazine published the following data concerning education level and attendance at Toronto Maple Leaf hockey games. A sample size of 955 was selected. Do the data indicate, at the 5% significance level, a relationship between level of education and attendance at Toronto Maple Leaf hockey games?

	ATTENDANCE		
EDUCATION	**More Than Once Per Year**	**No More Than Once Per Year**	**Total**
College graduate	82	121	**203**
Some college	77	128	**205**
High-school graduate	105	219	**324**
Not a high-school graduate	50	173	**223**
Total	**314**	**641**	**954**

Chap09\P9-2#4.xls

4. ProviSoir convenience stores in Quebec are open 24 hours a day. They are interested in knowing whether there is a relationship between the time of day and the size of purchase. One of the stores is selected at random. Store records are collected over a period of several weeks, and then 292 purchases are randomly selected. The information is summarized in the following table. Is there a relationship between time and size of purchase? Use $\alpha = 5\%$.

	SIZE OF PURCHASE			
TIME OF PURCHASE	**$2 or less**	**$2.01 to $7**	**Over $7**	**Total**
8 a.m. to 3:59 p.m.	63	37	12	**112**
4 p.m. to 11:59 p.m.	60	42	19	**121**
12 midnight to 7:59 a.m.	25	29	5	**59**
Total	**148**	**108**	**36**	**292**

Chap09\P9-2#5.xls

5. A Winnipeg pharmaceutical firm has surveyed a random sample of 135 people suffering from Parkinson's disease. Among the facts obtained is the following bivariate frequency distribution of disease duration and degree of self-reliance:

	DISEASE DURATION (years)			
SELF-RELIANCE	**Less than 5**	**5 to 9**	**10 to 14**	**15 or more**
Considerable	35	25	20	15
Little	5	10	10	15

Perform a hypothesis test to determine whether or not disease duration and degree of self-reliance are independent at the 1% level of significance.

CHAPTER 10
Analysis of Variance

10.1 ONE-WAY ANOVA *F*-TEST

ANOVA (analysis of variance) is a statistical test of significance for the equality of two or more population (treatment) means.

The following assumptions are made:

1. All populations are normally distributed.

2. All populations have the same variance: $\sigma_1^2 = \sigma_2^2 = \sigma_3^2 = \cdots = \sigma_r^2 = \sigma^2$.

3. Independent random samples are selected from the populations.

The ANOVA table is constructed as follows:

Source of Variation	Sum of Squares	df	Mean Square	*F**
Treatment	SSTR	$r-1$	MSTR = SSTR/$(r-1)$	MSTR/MSE
Error	SSE	$n_T - r$	MSE = SSE/$(n_T - r)$	
Total	SSTO	$n_T - 1$		

where n_T = the total number of observations
r = the number of treatments

Sum of Squares:

The total variability in the data may be attributed to two sources: (1) the treatment it received, and (2) residual (extraneous) factors. The first estimate (SSTR) is based on the differences *between* the treatment means and the overall mean. The second estimate (SSE) is based on the differences of observations *within* each treatment group from the corresponding treatment mean.

ANOVA hypothesis test:

$H_0: \mu_1 = \mu_2 = \mu_3 = \cdots = \mu_r$ (All the population means are equal.)
H_1: Not all μ's are equal. (At least one of the means is different.)

Test statistic: $F^* = \text{MSTR}/\text{MSE}$

Critical value: $F(1 - \alpha; r - 1; n_T - r)$

Decision rule: Conclude H_0 if $F^* \leq$ critical value.
Conclude H_1 if $F^* >$ critical value.

The ratio F^* serves as the test statistic for the ANOVA test of hypothesis. If the null hypothesis is true, then estimates of variance based on treatment (*between groups*) sum of squares (SSTR) and error (*within groups*) sum of squares (SSE) should be nearly of the same order of magnitude. The test statistic would be relatively small, which would lead to accepting the null hypothesis based on the above decision rule. On the other hand, if the alternative hypothesis is true, then MSTR should be expected to be significantly larger than MSE. The test statistic would be relatively large, which would lead to rejecting the null hypothesis and concluding that the treatments do have some effect.

If raw data are available, then the **Anova: Single Factor** option of the **Data Analysis** tool can be used. The steps for performing the hypothesis test are shown in the following example.

EXAMPLE: GMAT SCORES

Chap10\Ex10-1.xls

Consider the following data on GMAT scores of MBA applicants to three Canadian universities:

University of Montreal	University of Toronto	University of Alberta
560	630	690
640	620	680
580	510	570
720	540	510
420	520	620
	540	

Test the hypothesis that the mean GMAT scores are the same for the MBA applicants to the three universities, using a 5% level of significance.

EXAMPLE

Setting Up the Template

1. Type the headings and data above, or import the headings and data from Columns A, B, and C of file **Chap10\Ex10-1.xls**. Select Cells A1:C1. From the **Format** menu, choose **Column, Width**. Change the width to **20**. To format, select Cells A1:C1 and click **Bold** on the toolbar. Select Cells A1:C7 and click **Center** on the toolbar. If the data were in stacked format (all in one column), they would have to be unstacked (put in separate columns) in order to proceed.

2. From the **Tools** menu, choose **Data Analysis**. From the **Data Analysis** tools list, choose **Anova: Single Factor** and click **OK**.

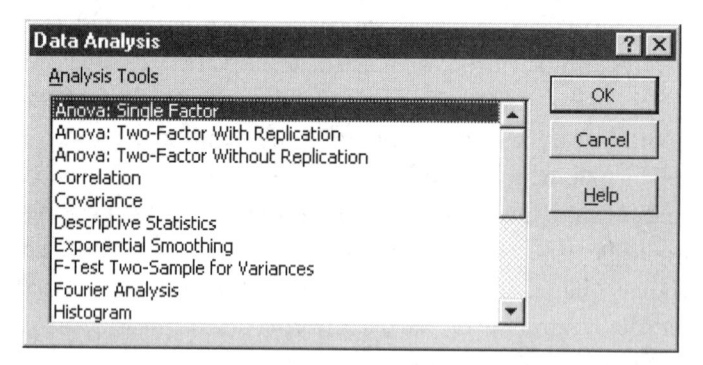

3. In the **Anova: Single Factor** dialog box, enter **A1:C7** for the **Input Range**. Choose the **Grouped By: Columns** option button, check the **Labels in First Row** check box, and enter **0.05** in the **Alpha** box. Choose the **Output Range** option button, then click in the box and enter **E1** for the **Output Range**. Click **OK**.

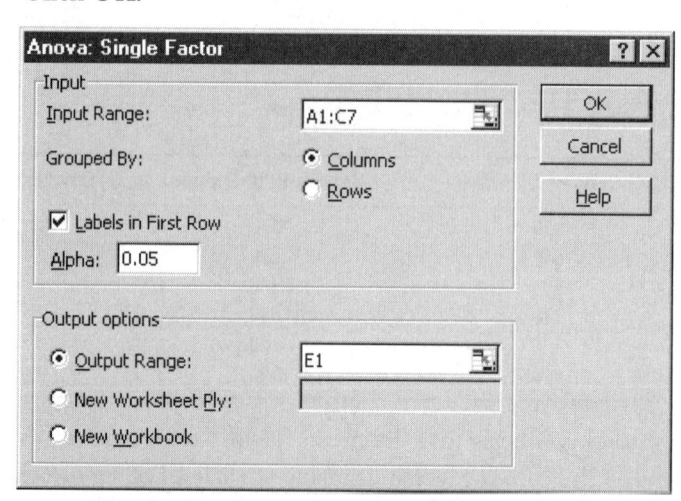

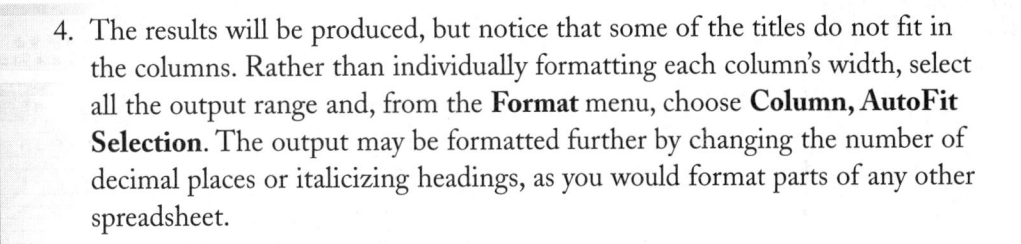

4. The results will be produced, but notice that some of the titles do not fit in the columns. Rather than individually formatting each column's width, select all the output range and, from the **Format** menu, choose **Column, AutoFit Selection**. The output may be formatted further by changing the number of decimal places or italicizing headings, as you would format parts of any other spreadsheet.

Discussing the Outcome

In the summary table, the sample size, total, mean, and variance are computed for each sample. The second table presents the ANOVA table along with the p-value and critical value for the hypothesis test. Since the ratio F^* (0.6051) is less than the critical value of 3.806, we cannot reject the null hypothesis. We therefore conclude that the mean GMAT scores are the same for MBA applicants to the three universities. Alternatively, the fact that the p-value of 0.56 is greater than the 0.05 level of significance leads to the same conclusion.

Using Excel, we can observe the effect of changes in the raw data on the ANOVA results. For example, change both the third and fourth values for the University of Alberta to 710, then use the **Anova: Single Factor** tool on the revised data set. The ratio F^* changes to 4.205, which is now greater than the critical value of 3.806; therefore, the null hypothesis is now rejected. We would conclude that not all the mean GMAT scores are the same for MBA applicants to the three universities. Alternatively, the fact that the new p-value of 0.039 is less than the 0.05 level of significance leads to the same conclusion. Note that whenever the raw data change, the **Data Analysis** tool must be used to get the updated ANOVA results. The output of the **Data Analysis** tool is *not* automatically updated to reflect changes in the data.

PROBLEMS

Chap10\P10-1#1.xls

1. It is suspected that four machines in a canning plant fill cans to different levels on the average. Random samples of cans were taken from each machine, and the fill in millilitres was measured. The following results were recorded:

Machine 1	Machine 2	Machine 3	Machine 4
323	329	325	320
322	326	324	323
320	327	322	321
322		325	

Do the machines appear to be filling the cans at different average levels at $\alpha = 1\%$?

Chap10\P10-1#2.xls

2. Most new products are test-marketed in several locations. Suppose the following table represents the number of sales for a new product at each of three locations during each of the last four months:

Red Deer, Alberta	Brampton, Ontario	Laval, Quebec
450	445	500
410	415	470
395	420	520
420	400	495

Determine whether there is a difference among the mean sales at the three locations, using $\alpha = 5\%$.

Chap10\P10-1#3.xls

3. The balances of randomly chosen clients from three major Canadian banks are shown below:

Royal Bank	Scotia Bank	CIBC
105	160	70
125	55	90
170	130	205
40	465	120
330	510	60
265		610
180		310
		135

Test at the 5% level of significance whether the mean balances are the same for the three Canadian banks.

PROBLEMS

4. The costs of computer maintenance for computers from three different university computer labs are shown below (in $1,000s):

McGill	Simon Fraser	Dalhousie
150	225	90
180	75	135
270	210	300
45	690	165
480	750	105
	390	930
	285	480
		195

Test the hypothesis $H_0: \mu_1 = \mu_2 = \mu_3$ with $\alpha = 1\%$.

5. CAE Inc. specializes in developing technology systems for a wide variety of customers such as commercial airlines, space agencies, the medical community, and the pulp and paper industry. CAE Inc. is headquartered in Canada and also has facilities throughout the United States, Europe, and Australia. The corporation employs 6200 people world-wide in a variety of disciplines and conducts a yearly study of its departmental salaries. The following table shows the yearly salaries of employees in four different departments:

Dept. #1	Dept. #2	Dept. #3	Dept. #4
$37,760	$44,863	$46,901	$38,845
$31,840	$36,728	$24,619	$27,912
$28,632	$28,418	$36,286	$33,215
$25,459	$35,220	$36,654	$36,476
$42,765	$55,306	$55,897	$47,014
$30,916	$34,821	$52,449	$35,234
$37,082	$35,345	$30,113	$27,870
$40,184	$25,490		$34,629
$27,677	$26,411		$33,586
$32,295			$42,994
$31,255			$38,704
$29,753			$34,773
			$35,166
			$67,139
			$70,376
			$64,528
			$47,652
			$50,913
			$45,200
			$32,639
			$53,622

Test at the 5% level of significance whether the mean yearly salaries are the same for employees of the four departments.

CHAPTER 11
Simple Linear Regression

11.1 INSERTING A TREND LINE AND THE TREND FUNCTION

Simple linear regression analysis is used to analyse the nature of the relationship between two variables. The dependent variable is designated by y and the independent variable is designated by x. The average value of the dependent variable is determined by the value of the independent variable. For a given independent variable, there is not just one possible value for the dependent variable. The relationship is estimated and then used to make predictions for the dependent variable.

This relationship is described by a straight line model in the general form

$$\textit{True regression line: } y = \beta_0 + \beta_1 x + \varepsilon$$
$$\textit{Estimated regression line: } \hat{y} = b_0 + b_1 x$$

where x = the independent variable
β_0 = the true y-intercept
β_1 = the true slope
ε = the error term
b_0 = the estimated y-intercept
b_1 = the estimated slope

The y-intercept is the point on the y-axis where the line crosses and is the average value of y when $x = 0$. The regression line either slopes upward (positive slope) or downward (negative slope), and the slope represents the average change in y when x is increased by 1. The y-intercept and the slope are called the *parameters* of the line.

SECTION 11.1 • INSERTING A TREND LINE AND THE TREND FUNCTION

Obviously many lines could be seen as fitting data in a scatter chart. The goal is to determine the line of best fit, which is the line that passes through the scatter of observations and deviates the least from the points. This line is called the *least squares regression line*, as it minimizes the sum of the squared deviations of the observations from the line. This is the amount of deviation not explained by the regression line.

The regression equation can then be used in predicting a value of y based on a given value of x by substituting the x-value into the regression line. Do not use the regression line to predict y with values of the independent variable beyond the range of those represented in the sample. The nature of the relationship outside the range of x may not be linear, and extrapolation may lead to false conclusions.

In Excel, the **Chart Wizard** must first be used to obtain a scatter chart for a set of data, as discussed in Section 1.4. The steps for then superimposing a regression line on the scatter chart and predicting a value of y based on a given value of x are shown in the following example.

EXAMPLE

PROFITS AND ADVERTISING EXPENDITURES Chap11\Ex11-1.xls

The VP of sales for the Zellers department store chain wishes to investigate the relationship between store profits (y, in $1,000s) and advertising expenditures (x, in $1,000s). The following data have been determined from a random sample of 10 stores:

Store	Advertising (x)	Profit (y)
1	3	9.4
2	3	10.3
3	4	10.9
4	4	9.9
5	5	12.9
6	5	11.8
7	6	11.5
8	6	13.2
9	7	12.8
10	7	12.1

Superimpose the regression line on the scatter chart for this data, and estimate the profit of a store that spends $3,500 on advertising.

EXAMPLE

Setting Up the Template

1. Type the "Advertising" and "Profit" headings and data above or import the headings and data from Columns A and B of file **Chap11\Ex11-1.xls**. Select Cells A1:B1. From the **Format** menu, choose **Column, Width**. Change the width to **15**. To format the headings, select Cells A1:B1 and click **Bold** on the toolbar.

2. Use the **Chart Wizard** to obtain the scatter chart as discussed in Section 1.4, "Setting Up the Template" (pp. 18–20).

3. Click anywhere within the chart box to put the chart into edit mode. From the **Chart** menu at the top of the screen, select **Add Trendline**. In the **Add Trendline** dialog box, select the **Type** tab and then choose **Linear** as the **Trend/Regression type**.

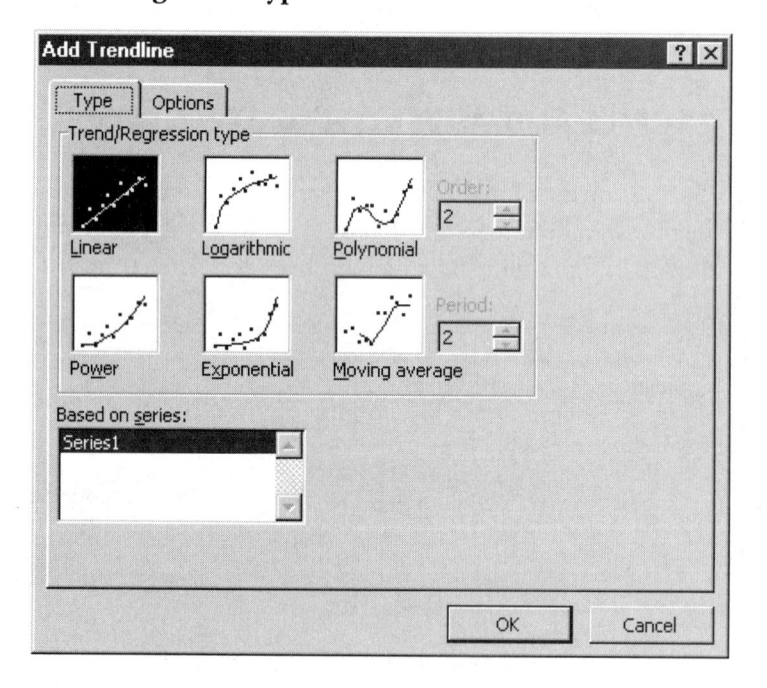

4. Click the **Options** tab and select the **Trendline name, Automatic** option button. Then select the **Display equation on chart** check box. Click the **OK** button.

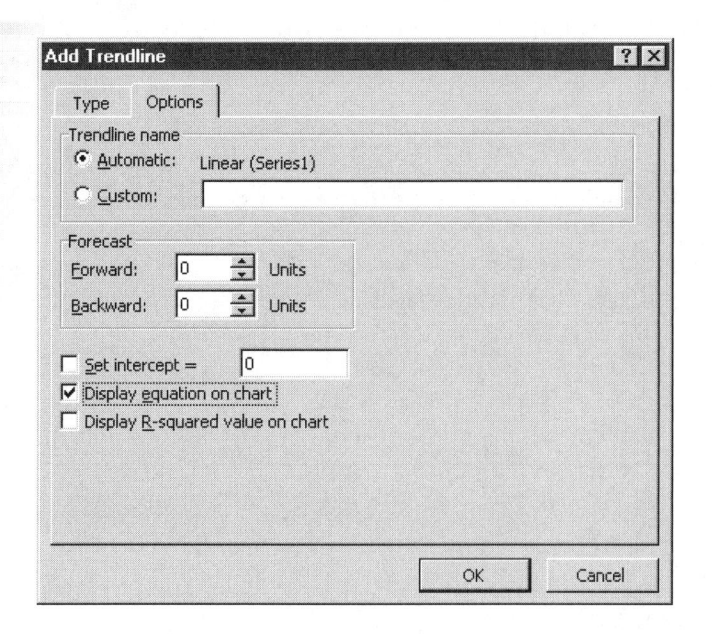

SECTION 11.1 • INSERTING A TREND LINE AND THE TREND FUNCTION

5. To predict the average value of *y* for a given value of *x*, use the **TREND** function. Enter the labels **Advertising** and **Predicted Profit** in Cells A14 and A15, respectively. In Cell B14, enter a value of **3.5** for the independent variable. In Cell B15, enter the formula **=TREND(B2:B11,A2:A11,B14)**. B2:B11 refers to the range of the dependent variable, and A2:A11 refers to the range of the independent variable.

Discussing the Outcome

The regression line and the regression equation of $y = 0.715x + 7.905$ appear on the chart. The profit of a store that spends $3,500 on advertising is therefore estimated to be $10,407.50.

Now that the template has been set up, we can observe the effect of changes in individual points on the fit of the regression line, as changes in the data will be immediately reflected in the fitted model. For example, change the profit for the first store in Cell B2 from 9.4 to 11.4. Notice how the regression line changes from $y = 0.715x + 7.905$ to $y = 0.515x + 9.105$.

You can reposition the regression equation by dragging it with the mouse. Double-click on the region containing the equation to call up the **Format Data Labels** dialog box to format the numbers. You can also use regular text editing options to rearrange the text and format the numbers. However, be aware that once you have used text editing, you can no longer increase and decrease the decimals in the dialog box, and changes in the data will no longer be reflected in the fitted model.

PROBLEMS

1. Refer to Problem 1 in Section 1.4 (p. 21). Superimpose the regression line on the scatter chart for this data, and estimate the number of faultless components produced by a worker with 30 weeks' experience.

2. Refer to Problem 2 in Section 1.4 (p. 22). Superimpose the regression line on the scatter chart for this data, and estimate the amount of home insurance carried by a person with an annual income of $50,000.

3. Refer to Problem 3 in Section 1.4 (p. 23). Superimpose the regression line on the scatter chart for this data, and estimate the consumer awareness (%) for a product with an advertising expenditure of $600,000.

11.2 USING THE DATA ANALYSIS TOOL FOR SIMPLE REGRESSION

Excel has a **Data Analysis** tool to perform a detailed regression analysis. The steps for setting up the simple regression summary output are shown in the following example.

EXAMPLE — PROFITS AND ADVERTISING EXPENDITURES   Chap11\Ex11-2.xls

Refer to the example in Section 11.1, which looked at the relationship between store profits (y, in $1,000s) and advertising expenditures (x, in $1,000s) for a random sample of 10 Zellers stores:

SECTION 11.2 • USING THE DATA ANALYSIS TOOL FOR SIMPLE REGRESSION ◀ 149

EXAMPLE

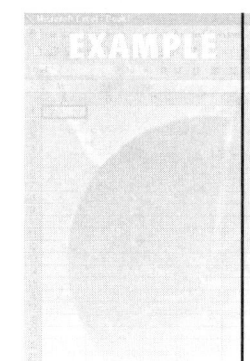

Store	Advertising (x)	Profit (y)
1	3	9.4
2	3	10.3
3	4	10.9
4	4	9.9
5	5	12.9
6	5	11.8
7	6	11.5
8	6	13.2
9	7	12.8
10	7	12.1

Produce the simple regression summary output.

Setting Up the Template

1. Type the "Advertising" and "Profit" headings and data above or import the headings and data from Columns A and B of file **Chap11\Ex11-2.xls**. Select Cells A1:B1. From the **Format** menu, choose **Column**, **Width**. Change the width to **15**. To format the headings, select Cells A1:B1 and click **Bold** on the toolbar.

2. From the **Tools** menu, choose **Data Analysis**. From the **Data Analysis** tools list, choose **Regression** and click **OK**.

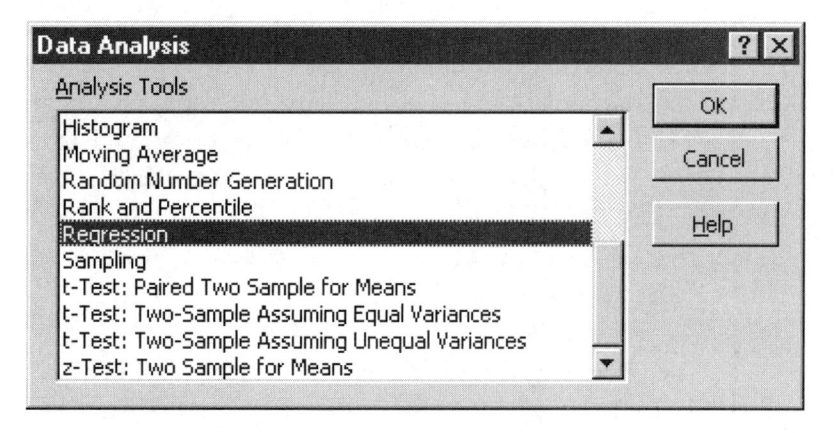

EXAMPLE

3. In the **Regression** dialog box, enter **B1:B11** for the **Input Y Range** and enter **A1:A11** for the **Input X Range**. Check the **Labels** check box. Choose the **Output Range** option button, then click in the box and enter **D1** for the **Output Range**. Click **OK**.

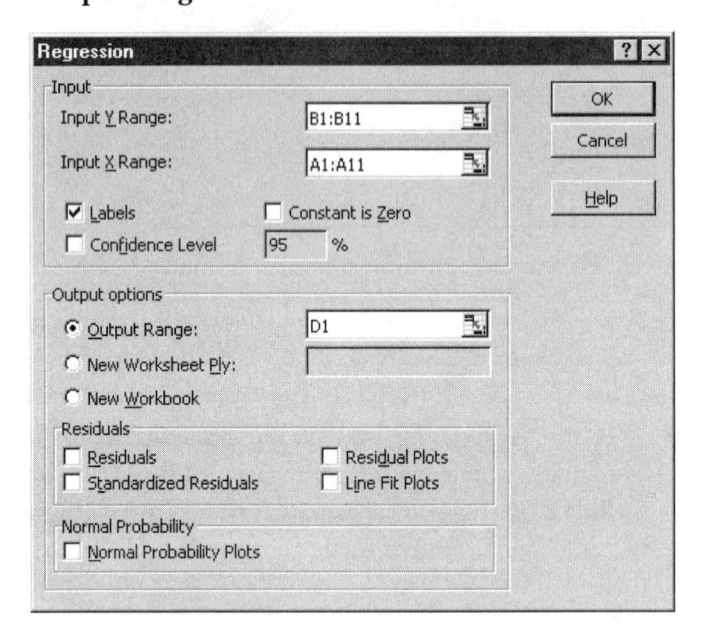

4. The results will be produced, but notice that some of the titles do not fit in the columns. Rather than individually formatting each column's width, select all the output range and, from the **Format** menu, choose **Column, AutoFit Selection**. The duplicate "Lower 95.0%" and "Upper 95.0%" at the far right of the output can be deleted. The output may be formatted further by changing the number of decimal places of numerical results or by italicizing headings, as you would format parts of any other spreadsheet.

Discussing the Outcome

The values for b_0 and b_1 can be found in Cells E17 and E18 of the third section of the summary output under the heading **Coefficients**. The estimated regression line is $\hat{y} = b_0 + b_1 x = 7.905 + 0.715x$. The value $b_1 = 0.715$ implies that, on the average, a 1-unit increase in x will cause y to increase by 0.715 (that is, if the advertising budget is increased by $1,000, the profit will increase, on average, by $715). The value $b_0 = 7.905$ implies that the average profit for stores with no advertising will be $7,905. *Caution:* $x = 0$ is not within the range of empirical data; beware the dangers of extrapolation.

EXAMPLE

Coefficient of Determination

The coefficient of determination in Cell E5 (**R Square**) measures the percent of variation in the dependent variable that is explained by the independent variable. In the example, 64.89% of the variation in store profit is explained by advertising. The coefficient of determination has the following properties:

1. r^2 is the square of the correlation coefficient and will always be between 0 and 1.

2. The closer r^2 is to 1, the better the regression model.

3. r^2 = SSR/SSTO.

Coefficient of Correlation

The coefficient of correlation in Cell E4 (**Multiple R**) measures the strength of the *linear* association between the independent and dependent variables. The coefficient of correlation has the following properties:

1. The value of r will always be between −1 and +1.

2. Positive correlation implies that x and y move in the same direction.

3. Negative correlation implies that x and y move in opposite directions.

4. The closer r is to +1, the stronger the direct linear relationship between x and y.

5. The closer r is to −1, the stronger the inverse linear relationship between x and y.

6. An r close to 0 indicates that there is little linear association between x and y.

7. r is the square root of r^2 and has the same sign as the slope of the regression line.

8. The closer r is to ±1, the more tightly the points are scattered around the regression line.

9. r = ±1 indicates a perfect linear association between x and y.

Linear Relationship

A linear relationship exists between the dependent variable and the independent variable only if the slope of the regression line is significantly different from 0. If $\beta_1 \neq 0$, then some positive or negative relationship does exist between x and y; if $\beta_1 = 0$, then knowing x does not help in predicting y. The decision rule for the hypothesis test is based on the p-value approach. The decision rule is the following:

$$\text{Do not reject } H_0 \text{ if } p\text{-value} \geq \alpha.$$
$$\text{Reject } H_0 \text{ if } p\text{-value} < \alpha.$$

Since the p-value of 0.0049 in Cell H18 is less than a 0.05 level of significance, we can conclude that there is a significant linear relationship between profit and advertising (that is, advertising should be kept in the model).

Confidence Interval Estimates

The 95% confidence interval estimates for the regression coefficients can be found in the third section of the summary output under the headings **Lower 95%** and **Upper 95%**.

The 95% confidence interval estimate for β_1 is as follows:

$$b_1 - t(1 - \alpha/2; n - 2)s(b_1) \leq \beta_1 \leq b_1 + t(1 - \alpha/2; n - 2)s(b_1)$$
$$0.2863 \leq \beta_1 \leq 1.1437$$

Since 0 is not contained in the above interval, then conclude $H_1: \beta_1 \neq 0$.

The 95% confidence interval estimate for β_0 is as follows:

$$b_0 - t(1 - \alpha/2; n - 2)s(b_0) \leq \beta_0 \leq b_0 + t(1 - \alpha/2; n - 2)s(b_0)$$
$$5.676 \leq \beta_0 \leq 10.134$$

Since 0 is not contained in the above interval, then conclude $H_1: \beta_0 \neq 0$.

Standard Error

The standard error of the estimate in Cell E7 measures the scatter of the actual profit values around the regression line. A lower standard error represents a closer fitted model.

Other results provided in the summary table, such as the **ANOVA** table and **Adjusted R Square**, are discussed in Chapter 12.

Note that whenever the raw data change, the **Data Analysis** tool must be used to get the updated results. The output of the **Data Analysis** tool is not automatically updated to reflect changes in the data.

PROBLEMS

1. Refer to Problem 1 in Section 1.4 (p. 21).
   a. What is the equation of the regression line? What is the meaning of the slope in the context of these data and the regression line?
   b. At the 5% level of significance, test for a significant linear relationship between x and y.
   c. What percentage of variation in number of faultless components is explained by the workers' experience?
   d. Find a 95% confidence interval for the slope of the true regression line.

2. Refer to Problem 2 in Section 1.4 (p. 22).
   a. What is the equation of the regression line? What is the meaning of the slope in the context of these data and the regression line?
   b. Estimate the amount of insurance carried for an income of $51,000.
   c. Find the coefficient of determination for these data and explain its meaning. What is the correlation coefficient?
   d. Test whether annual income is significant as a predictor for the amount of insurance carried. Use a 5% level of significance.

3. Refer to Problem 3 in Section 1.4 (p. 23).
   a. What is the equation of the regression line? What is the meaning of the slope in the context of these data and the regression line?
   b. Find the coefficient of determination for these data and explain its meaning. What is the correlation coefficient?
   c. What percentage of consumer awareness exists when advertising expenditure is known to be $470,000?
   d. Find a 95% confidence interval for the slope of the true regression line.

11.3 RESIDUAL ANALYSIS

A residual (e_i) is the difference between an actual y-value and the value predicted by the regression line for a given value of x. A simple technique to discover whether or not any of the regression assumptions are violated is to examine a plot of all the residuals against the independent variable x. If the scatter diagram of the residuals forms a horizontal band around 0, then there is no evidence that any of the assumptions have been violated. Other scatter diagrams of residuals could indicate a violation of the constant variance assumption, a curvilinear relationship rather than linear relationship between x and y, or a tendency of the residuals to be correlated and to move in a trend pattern.

Even with very good regression lines, there will always be some difference between the actual and the predicted values of y. This difference is the error term that appears in the equation for the population regression line:

$$y = \beta_0 + \beta_1 x + \varepsilon$$

Statistical inferences about a population regression relationship based on a sample regression relationship depend on certain assumptions made about the statistical characteristics of the error terms (residuals):

1. The error terms should be normally distributed with a mean of 0.

2. The variance of the error terms should be the same regardless of the given value of x.

3. The error terms should be statistically independent of each other.

4. The error terms should be statistically independent of x.

Excel has a **Data Analysis** tool to perform a residual analysis. The steps for setting up the residual analysis are shown in the following example.

EXAMPLE

PROFITS AND ADVERTISING EXPENDITURES Chap11\Ex11-3.xls

Refer to the example in Section 11.1, which looked at the relationship between store profits (y, in $1,000s) and advertising expenditures (x, in $1,000s) for a random sample of 10 Zellers stores:

Store	Advertising (x)	Profit (y)
1	3	9.4
2	3	10.3
3	4	10.9
4	4	9.9
5	5	12.9
6	5	11.8
7	6	11.5
8	6	13.2
9	7	12.8
10	7	12.1

Perform a residual analysis.

Setting Up the Template

1. Type the "Advertising" and "Profit" headings and data above or import the headings and data from Columns A and B of file **Chap11\Ex11-3.xls**. Select Cells A1:B1. From the **Format** menu, choose **Column, Width**. Change the width to **15**. To format the headings, select Cells A1:B1 and click **Bold** on the toolbar.

2. From the **Tools** menu, choose **Data Analysis**. From the **Data Analysis** tools list, choose **Regression** and click **OK**.

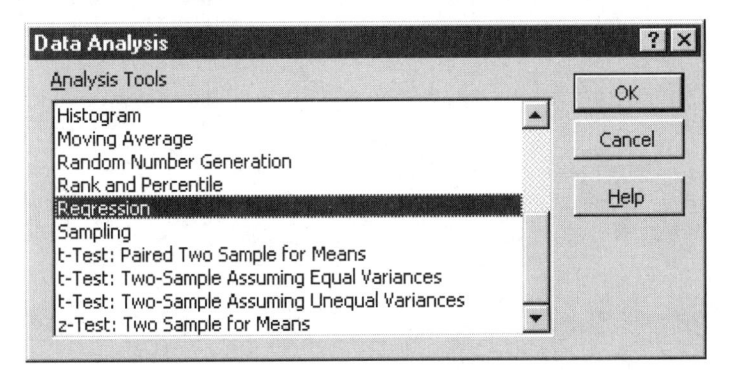

EXAMPLE

3. In the **Regression** dialog box, enter **B1:B11** for the **Input Y Range** and enter **A1:A11** for the **Input X Range**. Check the **Labels** check box. Choose the **Output Range** option button, then click in the box and enter **D1** for the **Output Range**. Check the **Residuals** and **Residual Plots** check boxes. Click **OK**.

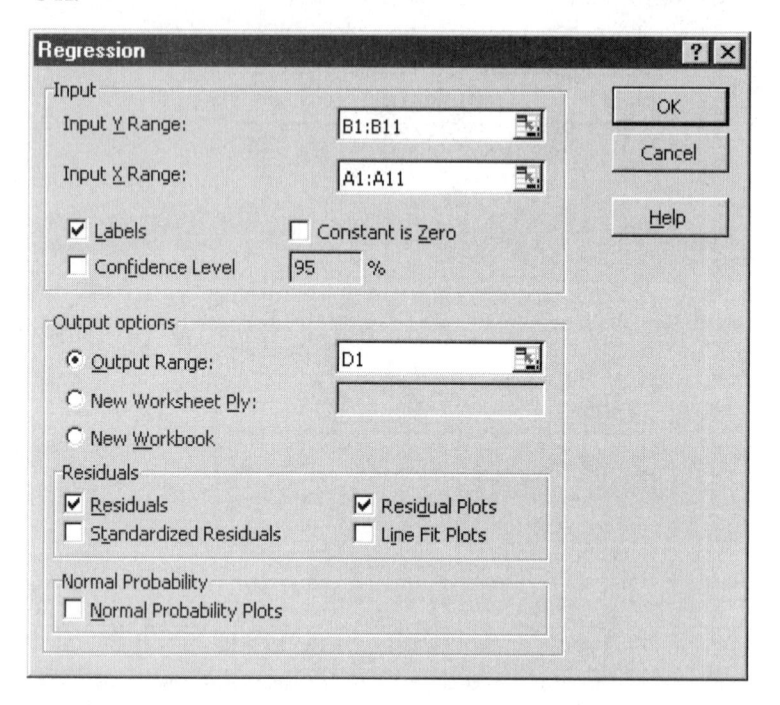

4. The results will be produced, but notice that some of the titles do not fit in the columns. Rather than individually formatting each column's width, select the output range and, from the **Format** menu, choose **Column**, **AutoFit Selection**. The duplicate "Lower 95.0%" and "Upper 95.0%" at the far right of the output can be deleted. The output may be formatted further by changing the number of decimal places of numerical results or by italicizing headings, as you would format parts of any other spreadsheet.

EXAMPLE

Discussing the Outcome

The predicted values of y, the residuals, and the residual plot are produced in addition to the regression summary output. A residual was previously defined as the difference between an actual y-value and the value predicted by the regression line for a given value of x. For example, the first store spent $3,000 on advertising. On average, one would expect this store to have a profit of $10,050, but its actual profit is $9,400. The residual for this store is $9,400 − $10,050 = −$650. The actual profit is $650 below what would be expected.

The model seems to be an adequate one, as the scatter diagram of the residuals forms a horizontal band around 0 and there is no evidence that any of the assumptions have been violated.

Note that whenever the raw data change, the **Data Analysis** tool must be used to get the updated results. The output of the **Data Analysis** tool is *not* automatically updated to reflect changes in the data.

PROBLEMS

1. Referring to Problem 1 in Section 1.4 (p. 21), perform a residual analysis.
2. Referring to Problem 2 in Section 1.4 (p. 22), perform a residual analysis.
3. Referring to Problem 3 in Section 1.4 (p. 23), perform a residual analysis.

11.4 OBTAINING CONFIDENCE INTERVALS FOR $E(y)$ AND PREDICTION INTERVALS FOR INDIVIDUAL RESPONSES

The confidence interval for the *mean* of the dependent variable for a given value of the independent variable is

$$E(y) = \hat{y} \pm t(1 - \alpha/2;\ n - 2)s(\hat{y})$$

where $s^2(\hat{y}) = \text{MSE}\left[\dfrac{1}{n} + \dfrac{(x-\bar{x})^2}{\Sigma(x-\bar{x})^2}\right]$

The prediction interval for a *single* value of the dependent variable for a given value of the independent variable is

$$y = \hat{y} \pm t(1 - \alpha/2;\ n-2)s(y)$$

where $s^2(y) = \text{MSE}\left[1 + \dfrac{1}{n} + \dfrac{(x-\bar{x})^2}{\Sigma(x-\bar{x})^2}\right]$

Excel does not have a **Data Analysis** tool to construct these intervals, so Excel functions and formulas can be used instead. The steps for setting up the confidence interval and prediction interval are shown in the following example.

EXAMPLE: PROFITS AND ADVERTISING EXPENDITURES — Chap11\Ex11-4.xls

Refer to the example in Section 11.1, which looked at the relationship between store profits (y, in $1,000s) and advertising expenditures (x, in $1,000s) for a random sample of 10 Zellers stores:

Store	Advertising (x)	Profit (y)
1	3	9.4
2	3	10.3
3	4	10.9
4	4	9.9
5	5	12.9
6	5	11.8
7	6	11.5
8	6	13.2
9	7	12.8
10	7	12.1

Construct a 95% confidence interval estimate for the mean profit of stores that spend $6,500 on advertising. Then construct a 95% prediction interval for the profit of a store that spends $6,500 on advertising.

Setting Up the Template

1. Type the "Advertising" and "Profit" headings and data above or import the headings and data from Columns A and B of file **Chap11\Ex11-4.xls**. Select Cells A1:B1. From the **Format** menu, choose **Column, Width**. Change the width to **15**. To format the headings, select Cells A1:B1 and click **Bold** on the toolbar.

2. Change the width of Column C to **15**, Column E to **25**, and Column F to **15**. Enter the labels from file **Chap11\Ex11-4.xls** in Cells C1 and E1:E16. To format, select Cells C1 and E1:E16 and click **Bold** on the toolbar.

3. To compute the values of $(x - \bar{x})^2$, enter the formula **=(A2-$F$5)^2** in Cell C2. Copy Cell C2 and paste to Cells C3:C11.

4. In Cells F1 and F2, enter the given value of the independent variable and the sample size, respectively. Using our example, enter **6.5** in Cell F1 and **10** in Cell F2.

5. In Cell F3, enter the formula **=F2-2** to compute the degrees of freedom.

6. In Cell F4, enter the function **=TINV(0.05,F3)** to compute the *t*-value.

7. In Cell F5, enter the function **=AVERAGE(A2:A11)** to compute the mean value of the independent variable.

8. In Cell F6, enter the function **=SUM(C2:C11)** to compute $\Sigma(x - \bar{x})^2$.

9. In Cell F7, enter the function **=SQRT(0.6914)** to compute the standard error of the estimate, which is the square root of MSE. The MSE value was in the ANOVA table produced in Section 11.2. If the value of MSE resides elsewhere in the spreadsheet, it can be cell-referenced in the **SQRT** function as opposed to just being entered as a value in the function.

10. In Cell F9, enter the function **=TREND(B2:B11,A2:A11,F1)** to compute the value of $\hat{y} = 7.905 + 0.715(6.5) = 12.5525$.

11. In Cell F10, enter the formula **=F4*F7*SQRT(1/F2+(F1-F5)^2/F6)** to compute the half-width of the confidence interval.

12. In Cell F11, enter the formula **=F9-F10** to compute the lower confidence bound.

13. In Cell F12, enter the formula **=F9+F10** to compute the upper confidence bound.

14. In Cell F14, enter the formula **=F4*F7*SQRT(1+1/F2+(F1-F5)^2/F6)** to compute the half-width of the prediction interval.

15. In Cell F15, enter the formula **=F9-F14** to compute the lower prediction bound.

16. In Cell F16, enter the formula **=F9+F14** to compute the upper prediction bound.

Discussing the Outcome

The confidence interval implies that the mean profit for stores that spend $6,500 on advertising lies between $11,668.60 and $13,436.40, with 95% confidence.

The prediction interval implies that the profit for an individual store that spends $6,500 on advertising lies between $10,441.12 and $14,663.88, with 95% confidence. Note that the prediction interval for a single value is *wider* than the confidence interval for the mean value, since individual variability is larger than mean variability.

Now that the template has been constructed, we can observe the effects of changes in certain values in the spreadsheet. For example, change the given value of the independent variable in Cell F1 from 6.5 to 5.5. Notice how the confidence bounds change to $11,194.37 and $12,480.63.

PROBLEMS

1. Refer to Problem 1 in Section 1.4 (p. 21). Estimate the mean number of faultless components produced by workers with 30 weeks' experience with a 95% confidence interval.

2. Refer to Problem 2 in Section 1.4 (p. 22). Estimate the amount of home insurance carried by a person with annual income of $50,000 with a 99% prediction interval.

3. Refer to Problem 3 in Section 1.4 (p. 23). Estimate the mean consumer awareness (%) for products with an advertising expenditure of $600,000.

CHAPTER 12
Multiple Regression

12.1 USING THE DATA ANALYSIS TOOL FOR MULTIPLE REGRESSION

Multiple regression is regression analysis with more than one independent variable. The regression line equations are as follow:

True regression line: $y = \beta_0 + \beta_1 x_1 + \beta_2 x_2 + \beta_3 x_3 + \cdots + \beta_k x_k + \varepsilon$
Estimated regression line: $\hat{y} = b_0 + b_1 x_1 + b_2 x_2 + b_3 x_3 + \cdots + b_k x_k$

where x_k = the independent variables
β_k = the true regression parameters
ε = the error term
b_k = the estimated regression coefficients = the change in y (the dependent variable) per unit increase in x_k with all other independent variables *held fixed*

Excel has a **Data Analysis** tool to perform a multiple regression analysis. The steps for setting up the multiple regression summary output are shown in the following example.

CHAPTER 12 • MULTIPLE REGRESSION

EXAMPLE: VARIATION IN STORE PROFITS

Chap12\Ex12-1.xls

Refer to the example in Section 11.1. The VP of sales decides to attempt to construct a regression model that will explain a higher percentage of the variation in store profits. A brainstorming session with middle management results in the addition of a *potential* explanatory (predictor) variable: x_2 = the number of in-store specials offered per day.

The data for the sample of 10 stores follow:

Store	Advertising (x_1)	Specials (x_2)	Profit (y)
1	3	2	9.4
2	3	4	10.3
3	4	5	10.9
4	4	1	9.9
5	5	5	12.9
6	5	5	11.8
7	6	3	11.5
8	6	4	13.2
9	7	5	12.8
10	7	3	12.1

Produce and interpret the multiple regression summary output for the data.

Setting Up the Template

1. Type the headings and data from the second, third, and fourth columns above or import the headings and data from Columns A, B, and C of file **Chap12\Ex12-1.xls**. Select Cells A1:C1. From the **Format** menu, choose **Column, Width**. Change the width to **15**. To format the headings, select Cells A1:C1 and click **Bold** and **Center** on the toolbar.

2. From the **Tools** menu, choose **Data Analysis**. From the **Data Analysis** tools list, choose **Regression** and click **OK**.

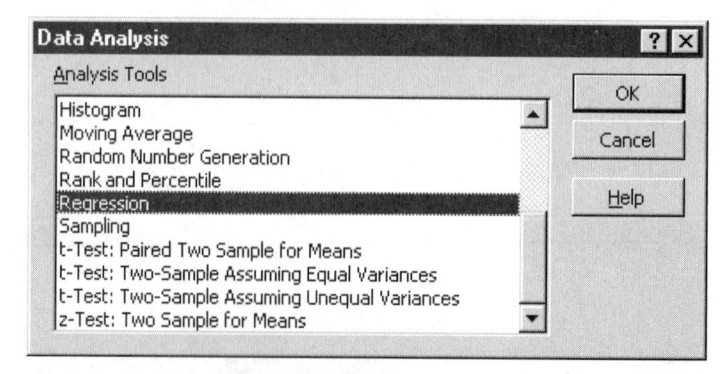

SECTION 12.1 • USING THE DATA ANALYSIS TOOL FOR MULTIPLE REGRESSION

3. In the **Regression** dialog box, enter **C1:C11** for the **Input Y Range** and enter **A1:B11** for the **Input X Range**. Check the **Labels** check box. Choose the **Output Range** option button, then click on the box and enter **E1** for the **Output Range**. Click **OK**.

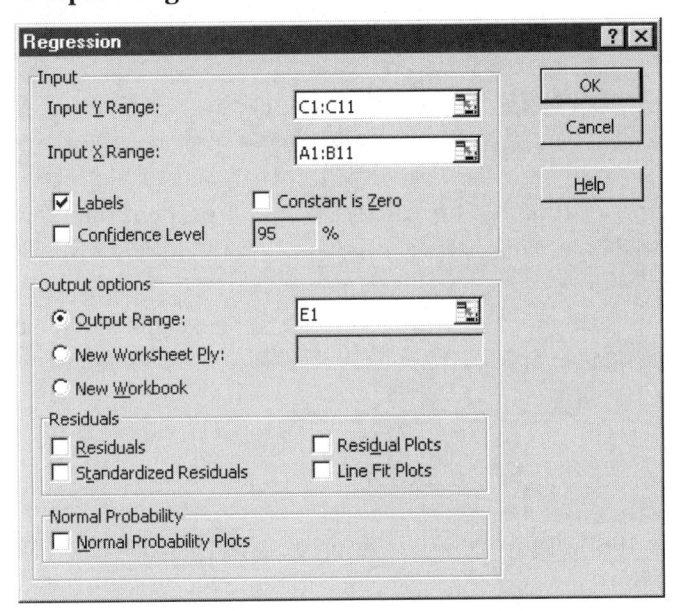

4. The results will be produced, but notice that some of the titles do not fit in the columns. Rather than individually formatting each column's width, select all the output range and, from the **Format** menu, choose **Column, AutoFit Selection**. The duplicate "Lower 95.0%" and "Upper 95.0%" at the far right of the output can be deleted. The output may be formatted further by changing the number of decimal places of numerical results or by italicizing headings, as you would format parts of any other spreadsheet.

Discussing the Outcome

The values for b_0, b_1, and b_2 can be found in Cells F17:F19 of the third section of the summary output under the heading **Coefficients**. The estimated regression line is $\hat{y} = 6.863 + 0.609x_1 + 0.425x_2$. The value $b_1 = 0.609$ implies that, on the average, a 1-unit increase in x_1 will cause y to increase by 0.609 (that is, if the advertising budget is increased by $1,000, the profit will increase, on average, by $609), provided x_2 remains fixed. The value $b_2 = 0.425$ implies that, on the average, a 1-unit increase in x_2 will cause y to increase by 0.425 (that is, if the number of in-store specials is increased by 1, the profit will increase, on average, by $425), provided x_1 remains fixed.

EXAMPLE

t-Test

t-tests are commonly used for testing the significance of an *independent variable*. The independent variable with the smallest *p*-value is the one that explains the most variation in *y*.

$$H_0: \beta_k = 0 \text{ (There is } no \text{ linear relationship between } x_k \text{ and } y.)$$
$$H_1: \beta_k \neq 0 \text{ (There } is \text{ a linear relationship between } x_k \text{ and } y.)$$

A linear relationship exists between the dependent variable and the independent variable only if its coefficient is significantly different from 0. If $\beta_k \neq 0$, then some positive or negative relationship exists between x and y; if $\beta_k = 0$, then knowing x does not help in predicting y. The decision rule for the hypothesis test is based on the *p*-value approach. The decision rule is the following:

Do not reject H_0 if *p*-value $\geq \alpha$.
Reject H_0 if *p*-value $< \alpha$.

Since the *p*-values for the two variables in Cells I18 and I19 are less than a 0.05 level of significance, we can conclude that both independent variables have significant relationships with store profit and therefore should be kept in the model.

Using the one-sided alternatives $H_1: \beta_k > 0$ and $H_1: \beta_k < 0$ tests for either a *direct* or *inverse* linear relationship between x_k and y respectively. Here the *p*-values would be divided by 2.

The 95% confidence interval estimates for the regression coefficients can be found in the third section of the summary output under the headings **Lower 95%** and **Upper 95%**.

The 95% confidence interval estimate for β_k is as follows:

$$b_k - t(1 - \alpha/2; n - p)s(b_k) \leq \beta_k \leq b_k + t(1 - \alpha/2; n - p)s(b_k)$$
$$0.2822 \leq \beta_1 \leq 0.9352$$
$$0.0820 \leq \beta_2 \leq 0.7684$$

Since 0 is not contained in the above intervals, we conclude $H_1: \beta_1 \neq 0$ and $H_1: \beta_2 \neq 0$.

EXAMPLE

F-Test

F-tests are commonly used for testing the overall significance of a *regression line*.

$$H_0: \beta_1 = \beta_2 = \beta_3 = \cdots = \beta_k = 0 \quad \text{(Regression line is \textit{not} significant.)}$$
$$H_1: \text{Not \textit{all} } \beta\text{'s} = 0 \quad \text{(Regression line \textit{is} significant.)}$$

ANOVA Table

Source of Variation	Sum of Squares	df	Mean Square	F*
Regression	SSR	$p - 1$	MSR = SSR/$(p - 1)$	MSR/MSE
Error	SSE	$n - p$	MSE = SSE/$(n - p)$	
Total	SSTO	$n - 1$		

where p = the number of regression parameters to be estimated

Since the *p*-value associated with the test statistic F^* in Cell J12 (0.0016) is less than a 0.05 level of significance, we can conclude that the regression line is significant.

Coefficient of Multiple Determination

The coefficient of multiple determination is the percent of variation in the dependent variable that is explained by the independent variables: r^2 = SSR/SSTO. In this example, 84.23% of the variation in store profit is explained by the two independent variables. This represents an improvement of 19.34% over the simple regression model in Section 11.2.

Adjusted R Square

Here r^2 is adjusted to take into account the sample size and the number of independent variables:

$$r^2_a = 1 - \frac{(n-1)}{(n-p)} (\text{SSE/SSTO})$$

The rationale behind adjustment is that if the number of independent variables is high relative to the sample size, the unadjusted r^2 value may appear to be unrealistically high. To avoid creating a false impression, the adjusted r^2 is often calculated. The adjusted value can decrease when a variable is added to the model if the decrease in $(n - p)$ is not offset by the decrease in SSE. If the number of independent variables is large, the actual and adjusted r^2 values will differ substantially. In this example, r^2_a in Cell F6 is 79.72%.

Standard Error

The standard error of the estimate of 0.5958 in Cell F7 represents an improvement over the 0.8315 standard error from the simple regression model in Section 11.2.

Note that whenever the raw data change, the **Data Analysis** tool must be used to get the updated results. The output of the **Data Analysis** tool is not automatically updated to reflect changes in the data.

PROBLEMS

Chap12\P12-1#1.xls

1. A large real-estate firm has developed an interest in selling houses in a particular area of London, Ontario. This area is undergoing urban development and has for sale a large number of older homes that will require a significant amount of repair and renovation. In a study to determine the relationship, if any, between the selling price of houses (PRICE), the number of bedrooms (BEDS), and the estimate of necessary repair costs (COST), the following data have been collected:

PRICE ($)	BEDS	COST ($100s)
109,000	2	80
145,400	3	70
129,700	2	35
135,200	2	30
134,500	3	40
127,100	3	70
128,200	2	85
148,500	4	70
146,000	5	50
170,900	6	50
166,700	6	45
159,300	4	35
160,500	6	55
156,400	6	60
168,100	5	35

PROBLEMS

a. Perform a multiple regression analysis of PRICE on the independent variables BEDS and COST.

b. Obtain a point estimate for the selling price of a three-bedroom house with an estimated cost of renovations of $7,500.

c. What percentage of variation in PRICE is "explained" by the combined effect of the two independent variables?

d. Rank the three independent variables according to their importance in the model. Are they all significant? If not, what action would you recommend?

Chap12\P12-1#2.xls

2. The following table shows data for 10 Manitoba crown corporations:

Crown Corporation	Gross Revenue ($1,000s)	Assets ($1,000s)	Employees
Canadian Wheat Board	$3,873,441	$8,858,583	480
Manitoba Hydro	$950,400	$5,900,000	4,000
Western Canada Lottery Corporation	$612,071	$44,028	175
Manitoba Telephone System	$530,787	$1,497,597	4,408
Manitoba Public Insurance	$383,194	$846,499	1,085
Manitoba Lotteries Corporation	$342,894	$74,913	778
Manitoba Liquor Control Commission	$340,652	$14,472	485
Manitoba Crop Insurance Corporation	$189,400	$190,800	88
Workers Compensation Board of Manitoba	$172,319	$500,239	380
The City of Winnipeg Hydro Electric System	$119,340	$135,179	594

Source: Manitoba Business Magazine, July/August 1995.

Discuss the results of the multiple regression analysis of Gross Revenue on the independent variables Assets and Employees.

Chap12\P12-1#3.xls

3. In order to predict the time required to package product, the manager of a mail-order company takes a random sample of 10 parcels and wishes to perform a multiple regression of the time to package the parcel, y = TIME (in minutes), against the independent variables x_1 = VOLUME (in cubic inches) and x_2 = WEIGHT (in pounds). The sample data collected follow:

PROBLEMS

PARCEL	TIME	VOLUME	WEIGHT
1	4.6	110	7.1
2	8.7	125	9.9
3	2.1	85	4.5
4	5.8	115	7.5
5	6.2	75	9.1
6	7.3	135	10.0
7	1.4	70	2.7
8	5.0	90	7.3
9	2.6	75	4.1
10	7.2	105	9.0

a. Perform a multiple regression analysis of TIME on the independent variables VOLUME and WEIGHT.

b. What percentage of variation in TIME is "explained" by the combined effect of the two independent variables?

c. Rank the two independent variables according to their importance in the model. Are they both significant? If not, what action would you recommend?

d. Perform the hypothesis test $H_0: \beta_1 = \beta_2 = 0$ at the 5% level of significance.

e. Determine the adjusted coefficient of multiple determination.

12.2 USING THE DATA ANALYSIS TOOL FOR CORRELATION

Correlation coefficients have the following properties:

1. The value of a correlation coefficient will always be between −1 and +1, and it measures the strength of the *linear* relationship between two variables.

2. Positive correlation implies that two variables move in the same direction.

3. Negative correlation implies that two variables move in opposite directions.

4. The closer a correlation is to +1, the stronger the *direct* linear relationship between two variables.

5. The closer a correlation is to −1, the stronger the *inverse* linear relationship between two variables.

6. A correlation close to 0 indicates that there is little linear association between two variables.

7. A correlation equal to ±1 indicates a perfect linear association between two variables.

Multicollinearity occurs when the variability of an independent variable plays a larger role in explaining the variability in another independent variable than in explaining the variability in the dependent variable. This does not mean that, on its own, the independent variable does not explain a significant amount of variation in the dependent variable. It simply means that there are variables in the model that perform the same job in explaining the variation in the dependent variable; therefore, they are not all needed in the regression model together. Multicollinearity can be detected in the following ways:

1. There exists a high correlation between the independent variables.

2. The estimated regression coefficients change when a variable is added or deleted.

3. There are conflicting results between the *F*-test and the *t*-tests.

4. There are estimated regression coefficients with a sign that is opposite of that expected.

5. The standard errors of the regression coefficients are inflated.

Multicollinearity is a problem because estimated regression coefficients are rendered meaningless, and inferential statistics (that is, *t*-tests, confidence intervals for β_k) are not reliable. However, in the case of multicollinearity, R^2 and the predictive power of the line remain unaffected.

Excel has a **Data Analysis** tool to perform a correlation analysis. The steps for producing the correlation matrix for a set of variables are shown in the following example.

EXAMPLE: VARIATION IN STORE PROFITS

Chap12\Ex12-2.xls

Refer to the example in Section 12.1. The VP of sales decides to examine the relationship between the variables of advertising, specials, and profit.

The data for the sample of 10 stores are as follow:

Store	Advertising (x_1)	Specials (x_2)	Profit (y)
1	3	2	9.4
2	3	4	10.3
3	4	5	10.9
4	4	1	9.9
5	5	5	12.9
6	5	5	11.8
7	6	3	11.5
8	6	4	13.2
9	7	5	12.8
10	7	3	12.1

Produce and interpret the correlation matrix for the data.

Setting Up the Template

1. Type the headings and data from the second, third, and fourth columns above or import the headings and data from Columns A, B, and C of file **Chap12\Ex12-2.xls**. Select Cells A1:C1. From the **Format** menu, choose **Column, Width**. Change the width to **15**. To format the headings, select Cells A1:C1 and click **Bold** and **Center** on the toolbar.

2. From the **Tools** menu, choose **Data Analysis**. From the **Data Analysis** tools list, choose **Correlation** and click **OK**.

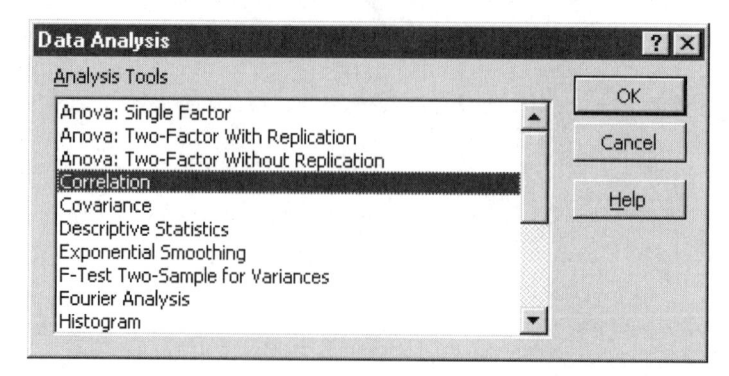

EXAMPLE

3. In the **Correlation** dialog box, enter **A1:C11** for the **Input Range**. Check the **Labels in First Row** check box. Choose the **Output Range** option button, then click in the box and enter **E1** for the **Output Range**. Click **OK**.

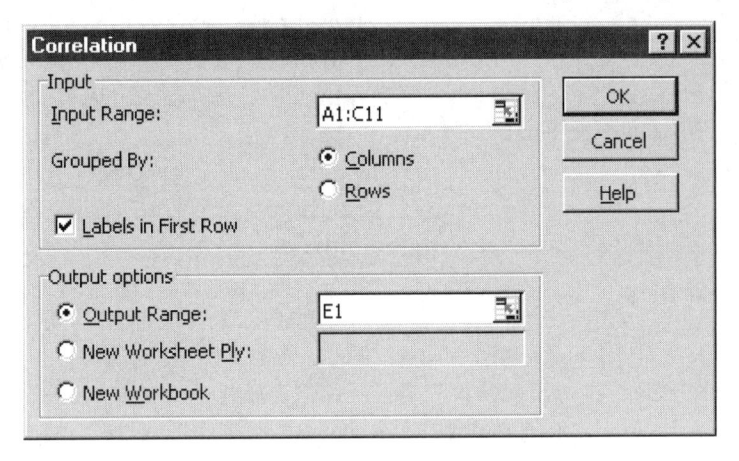

4. The results will be produced, but notice that some of the titles do not fit in the columns. Rather than individually formatting each column's width, select all the output range and, from the **Format** menu, choose **Column, AutoFit Selection**. The output may be formatted further by changing the number of decimal places of numerical results or by italicizing headings, as you would format parts of any other spreadsheet.

Discussing the Outcome

The correlation matrix displays the individual correlation coefficients between all possible pairs of variables. In our example, the correlation between profit and advertising is 0.8056. This implies that there is a strong positive linear relationship between profit and advertising. The correlation between profit and specials is 0.6360. This implies that there is a moderately strong positive linear relationship between profit and specials. Although both advertising and specials are useful variables in predicting profit, advertising is the better explanatory variable since it has the higher correlation coefficient.

The correlation between advertising and specials is 0.2628. This implies that there is a fairly weak positive relationship between advertising and specials and therefore multicollinearity in a multiple regression model to predict profit should not be a problem. Note that a correlation that is close to 0 does not necessarily imply that there is *no* relationship between x and y but simply that there is a very weak *linear* relationship between *x* and *y*. For instance, if there were a strong quadratic relationship between *x* and *y*, one would expect the correlation coefficient to be close to 0.

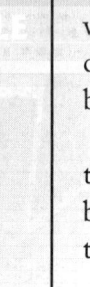

The diagonal values of 1 represent the fact the each variable has a perfect positive correlation with itself. The upper right-hand portion of the matrix is blank because, since the matrix is symmetrical, the values would be the same as those in the lower left-hand portion of the matrix.

Note that whenever the raw data change, the **Data Analysis** tool must be used to get the updated results. The output of the **Data Analysis** tool is *not* automatically updated to reflect changes in the data.

PROBLEMS

1. Referring to Problem 1 in Section 12.1 (p. 166), produce and interpret a correlation matrix.

2. Referring to Problem 2 in Section 12.1 (p. 167), produce and interpret a correlation matrix.

3. Referring to Problem 3 in Section 12.1 (p. 167), produce and interpret a correlation matrix.

12.3 INDICATOR VARIABLES

An *indicator variable* (or dummy variable) is a variable that can assume either one of only two values (0 or 1), where one value indicates the existence of a certain condition and the other value indicates that the condition does not hold. In general, to represent n possible conditions, we must create $n - 1$ indicator variables. For example:

$$\hat{y} = b_0 + b_1 x_1 + b_2 x_2 + b_3 x_3 + b_4 x_4$$

where x_1 = some quantitative variable
x_2 = 1 (if the season is winter); x_2 = 0 (if not)
x_3 = 1 (if the season is spring); x_3 = 0 (if not)
x_4 = 1 (if the season is summer); x_4 = 0 (if not)

Therefore, $x_2 = 0$, $x_3 = 1$, $x_4 = 0$ would indicate spring, and $x_2 = 0$, $x_3 = 0$, $x_4 = 0$ would indicate fall.

Excel's **Data Analysis** tool for regression can be used to analyse data that include indicator variables. The steps for setting up the analysis are shown in the following example.

EXAMPLE

VARIATION IN PROFITS BASED ON LOCATION Chap12\Ex12-3.xls

Refer to the example in Section 12.1. The VP of sales for Zellers decides that, in addition to advertising and specials, he should consider the location of stores in trying to explain variation in profits. The stores in the sample are either in Montreal ($x_3 = 1$) or in Toronto ($x_3 = 0$). The data for the sample of 10 stores follow:

Store	Advertising (x_1)	Specials (x_2)	Place (x_3)	Profit (y)
1	3	2	1	9.4
2	3	4	1	10.3
3	4	5	1	10.9
4	4	1	1	9.9
5	5	5	1	12.9
6	5	5	0	11.8
7	6	3	0	11.5
8	6	4	0	13.2
9	7	5	0	12.8
10	7	3	0	12.1

Test for a difference in profit between stores in Montreal and Toronto, using H_0: $\beta_3 = 0$ versus H_1: $\beta_3 \neq 0$ ($\hat{y} = b_0 + b_1 x_1 + b_2 x_2 + b_3 x_3$).

Setting Up the Template

1. Type the headings and data from the second through fifth columns above or import the data from Columns A through D of file **Chap12\Ex12-3.xls**. Select Cells A1:D1. From the **Format** menu, choose **Column, Width**. Change the width to **15**. To format the headings, select Cells A1:D1 and click **Bold** and **Center** on the toolbar.

2. From the **Tools** menu, choose **Data Analysis**. From the **Data Analysis** tools list, choose **Regression** and click **OK**.

CHAPTER 12 • MULTIPLE REGRESSION

EXAMPLE

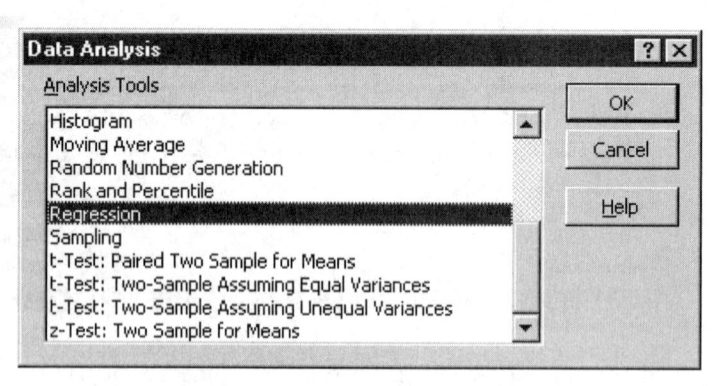

3. In the **Regression** dialog box, enter **D1:D11** for the **Input Y Range** and enter **A1:C11** for the **Input X Range**. Check the **Labels** check box. Choose the **Output Range** option button, then click in the box and enter **F1** for the **Output Range** and click **OK**.

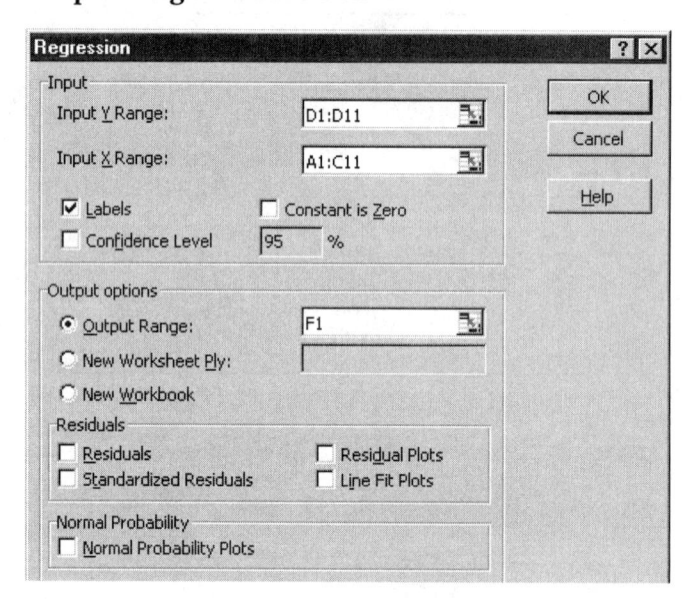

4. The results will be produced, but notice that some of the titles do not fit in the columns. Rather than individually formatting each column's width, select all the output range and, from the **Format** menu, choose **Column, AutoFit Selection**. The duplicate "Lower 95.0%" and "Upper 95.0%" at the far right of the output can be deleted. The output may be formatted further by changing the number of decimal places of numerical results or by italicizing headings, as you would format parts of any other spreadsheet.

EXAMPLE

Discussing the Outcome

Since the *p*-values for the two quantitative variables in Cells J18 and J19 are less than a 0.05 level of significance, we can conclude that both independent variables, x_1 and x_2, have significant relationships with store profit and therefore should be kept in the model.

If the indicator variable is a significant variable, then its coefficient b_3 represents the difference in profit between stores in Montreal and Toronto:

If $x_3 = 1$ (the store is in Montreal), then $\hat{y} = b_0 + b_1 x_1 + b_2 x_2 + b_3$.
If $x_3 = 0$ (the store is in Toronto), then $\hat{y} = b_0 + b_1 x_1 + b_2 x_2$.

If b_3 is positive, then stores in Montreal make a higher profit, on average; if b_3 is negative, then stores in Toronto make a higher profit, on average.

In our example, $b_3 = +0.4143$ (Cell G20). However, note that the *p*-value for the indicator variable in Cell J20 is greater than a 0.05 level of significance, so we can conclude that there is no evidence of a difference in store profit between stores in Montreal and Toronto.

OTHER TYPES OF REGRESSION MODELS

Interaction:

$$\hat{y} = b_0 + b_1 x_1 + b_2 x_2 + b_3 x_1 x_2$$

The effect of x_2 on y depends on the level of x_1 and vice versa.

To test for interaction: $H_0: \beta_3 = 0$ versus $H_1: \beta_3 \neq 0$

Quadratic Regression:

$$\hat{y} = b_0 + b_1 x_1 + b_2 x_2^2$$

The relationship between x_2 and y is a quadratic relationship.

To test for a quadratic relationship: $H_0: \beta_2 = 0$ versus $H_1: \beta_2 \neq 0$

PROBLEMS

1. Refer to Problem 1 in Section 12.1 (p. 166). Add the following indicator variable to the model: $x_3 = 1$ (if the house is close to a school); $x_3 = 0$ (if the house is not close to a school).

   x_3: 1, 1, 0, 0, 1, 0, 1, 0, 1, 1, 0, 0, 0, 1, 0

   Test for a difference in house prices between houses close to a school and houses not close to a school.

2. Refer to Problem 3 in Section 12.1 (p. 167). Add the following indicator variable to the model: $x_3 = 1$ (if the parcel is glassware); $x_3 = 0$ (if the parcel is not glassware).

   x_3: 1, 0, 1, 1, 1, 0, 0, 0, 1, 0

   Test for a difference in the time required to package a parcel between parcels that are glassware and parcels that are not glassware.

3. Refer to the example in this section. Test for an interaction effect between advertising and in-store specials.

CHAPTER 13
Time-Series Analysis and Forecasting

13.1 MOVING AVERAGES

A moving average of order L is a series of means computed over consecutive sequences of L observations. If L is an odd number, then moving averages are centred at the middle periods of the consecutive sequences. If L is an even number, then moving averages are centred between the two middle periods of the consecutive sequences. No moving averages can be computed for the first or last $(L - 1)/2$ periods. Moving averages are used to "smooth" data and eliminate cyclical fluctuations. They provide trend information that a simple average of historical data would hide.

Excel has a **Data Analysis** tool to compute moving averages. The steps for producing a set of moving averages for a time-series variable are shown in the following example.

EXAMPLE — WATER TEMPERATURE BY HOUR
Chap13\Ex13-1.xls

Based on concerns voiced by a local citizens' group, Environment Canada wishes to study the impact of a manufacturing plant built on the Peace River in northeastern Alberta. The citizens' group claims that the plant's activities are raising the temperature of the water in the river and thereby damaging the delicate aquatic ecosystems that have existed for thousands of years. In response, Environment Canada has measured the water temperature hourly for four consecutive fall days at an area near the plant that is believed to be most affected. The following table gives sample temperature measurements in degrees Celsius, beginning at 12:00 a.m.:

EXAMPLE

Hour	Temp
1	−11.54
2	−11.96
3	−13.37
4	−13.27
5	−13.62
6	−14.03
7	−14.36
8	−14.32
9	−14.45
10	−14.23
11	−14.50
12	−11.71
⋮	⋮

(complete data on disk)

Compute three-period moving averages for the data above.

Setting Up the Template

1. Import the headings and data from Columns A and B of file **Chap13\Ex13-1.xls**. Select Cells A1:C1. From the **Format** menu, choose **Column**, **Width**. Change the width to **10**. To format the headings, select Cells A1:B1 and click **Bold** and **Center** on the toolbar.

2. From the **Tools** menu, choose **Data Analysis**. From the **Data Analysis** tools list, choose **Moving Average** and click **OK**.

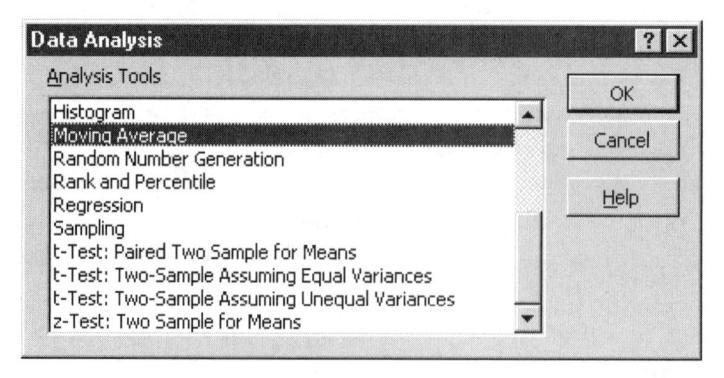

EXAMPLE

3. In the **Moving Average** dialog box, enter **B1:B97** for the **Input Range**. Check the **Labels in First Row** check box, and enter **3** for **Interval**. Choose the **Output Range** option button, then click in the box and enter **C1:C97** for the **Output Range**. Click **OK**.

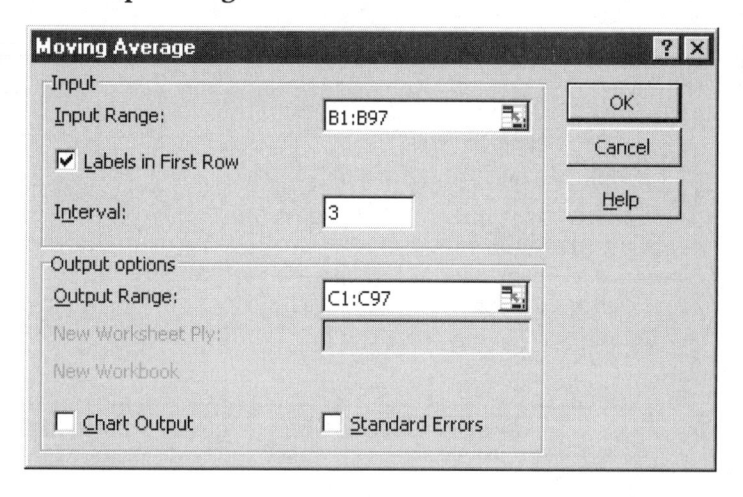

```
Moving Average                                    ? X
┌─Input──────────────────────────────┐       ┌──────────┐
│ Input Range:          [B1:B97]    │       │    OK    │
│ ☑ Labels in First Row             │       ├──────────┤
│                                    │       │  Cancel  │
│ Interval:             [3      ]   │       ├──────────┤
│                                    │       │   Help   │
├─Output options─────────────────────┤       └──────────┘
│ Output Range:         [C1:C97]    │
│ New Worksheet Ply:    [       ]   │
│ New Workbook                       │
│                                    │
│ ☐ Chart Output    ☐ Standard Errors│
└────────────────────────────────────┘
```

4. The output may be formatted further by changing the number of decimal places in Cells C3:C96 to 2. Cut the **#N/A** from Cell C1 and paste to Cell C97. In Cell C1, add the heading **Moving Averages**, format it by clicking **Bold** and **Center** on the toolbar, and adjust the column width by choosing **AutoFit Selection** under **Format, Column**.

Discussing the Outcome

The average of the first series of three observations (–11.54, –11.96, –13.37) is –12.29, the average of the next series of three observations (–11.96, –13.37, –13.27) is –12.87, and so on.

Note that whenever the raw data change, the output of the **Data Analysis: Moving Average** tool is automatically updated to reflect changes in the data, as **=AVERAGE** functions have been placed in Cells C3:C96. For example, change the temperature for Hour 1 in Cell B2 from –11.54 to –11.04. Notice how the moving average in Cell C3 changes from –12.29 to –12.12.

PROBLEMS

1. The following data show the number of users (in 1,000s) of a Sears credit card during the period 1980 to 1996:

Year	1980	1981	1982	1983	1984	1985	1986	1987	1988
Users	3.91	3.89	4.20	4.24	4.22	4.27	4.53	4.58	4.66

Year	1989	1990	1991	1992	1993	1994	1995	1996
Users	4.62	4.67	4.75	4.85	4.80	4.86	4.91	4.94

Compute three-period moving averages for the number of users.

2. Consider the following data for quarterly sales of a cough medicine (in $1,000s) in a Manitoba over a seven-year period from 1991 to 1997:

	Quarter 1	Quarter 2	Quarter 3	Quarter 4
1991	12.5	14.9	18.7	16.1
1992	13.5	16.0	19.6	16.8
1993	14.7	17.1	20.4	18.2
1994	15.4	18.6	21.5	19.3
1995	16.3	19.5	21.6	20.0
1996	17.2	20.8	23.2	21.8
1997	18.0	21.9	24.1	22.4

Compute four-period moving averages for the quarterly sales.

3. The following data show the quarterly sales of a brewing company (in $1,000,000s) during the period 1993 to 1997:

Quarter	1993	1994	1995	1996	1997
1	166	161	185	178	175
2	220	217	211	243	234
3	311	314	312	298	297
4	225	216	227	226	213

Compute four-period moving averages for the quarterly sales.

13.2 EXPONENTIAL SMOOTHING

Exponential smoothing is another way to "smooth" data and eliminate cyclical fluctuations, thereby giving a better indication of the long-term trend of the time series. It provides forecasts one period into the future, and forecasts are made each period for succeeding periods.

Exponential smoothing weights more recent observations more heavily than remote observations. Older observations are given successively smaller weights; therefore, each forecast is dependent on all previous observations, unlike the method of moving averages:

$$\text{Forecast}(t + 1) = ay_t + (1 - a) \times \text{Forecast}(t)$$

where a is called the smoothing constant and is a value between 0 and 1. The higher the smoothing constant is, the faster past observations are dampened out and the more responsive the smoothing procedure is to changes in the average level of the time series. The lower the smoothing constant is, the slower past observations are dampened out and the less responsive the smoothing procedure is to changes in the average level of the time series. The more the average level is changing, the larger the smoothing constant should be. Although the selection of the smoothing constant is rather subjective, it should generally not be greater than 0.3, because if the average level of the time series is changing quite quickly over time, the exponential smoothing technique will not be effective in producing forecasts. Exponential smoothing is more appropriate when the average level of the time series is changing slowly over time.

Excel has a **Data Analysis** tool for exponential smoothing. The steps for applying exponential smoothing are shown in the following example.

EXAMPLE — WATER TEMPERATURE BY HOUR   Chap13\Ex13-2.xls

Refer to the example in Section 13.1, in which Environment Canada measured the water temperature hourly for four consecutive fall days at an area of the Peace River believed to be affected by a manufacturing plant. The following table gives sample temperature measurements in degrees Celsius, beginning at 12:00 a.m.:

EXAMPLE

Hour	Temp
1	−11.54
2	−11.96
3	−13.37
4	−13.27
5	−13.62
6	−14.03
7	−14.36
8	−14.32
9	−14.45
10	−14.23
11	−14.50
12	−11.71
⋮	⋮

(complete data on disk)

Smooth the time series using exponential smoothing with a smoothing constant of 0.2.

Setting Up the Template

1. Import the headings and data from Columns A and B of file **Chap13\Ex13-2.xls.** Select Cells A1:C1. From the **Format** menu, choose **Column, Width.** Change the width to **10.** To format the headings, select Cells A1:B1 and click **Bold** and **Center** on the toolbar.

2. From the **Tools** menu, choose **Data Analysis.** From the **Data Analysis** tools list, choose **Exponential Smoothing** and click **OK.**

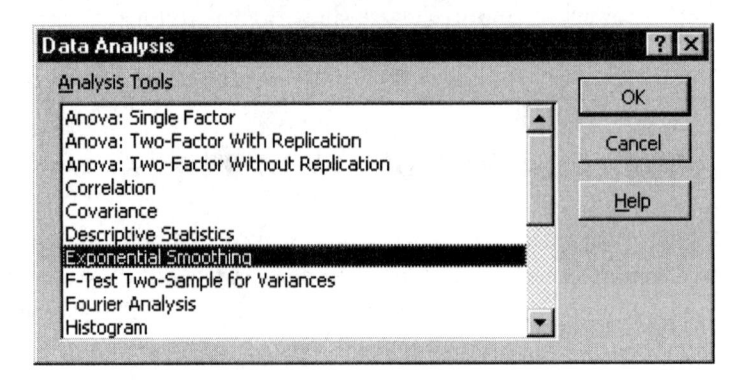

EXAMPLE

3. In the **Exponential Smoothing** dialog box, enter **B1:B97** for the **Input Range**. Check the **Labels** check box, and enter **0.8** (1 − smoothing constant) for the **Damping Factor**. Enter **C2:C97** for the **Output Range** and click **OK**.

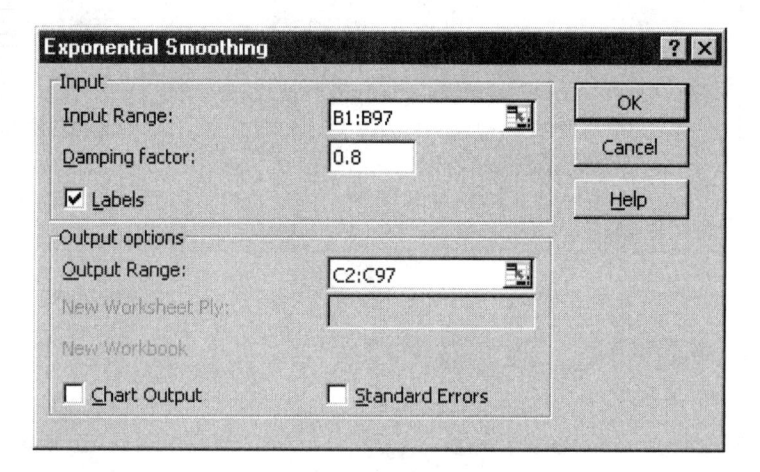

4. Enter the title **ES (a = 0.2)** in Cell C1. Cut Cells C3:C97 and paste to Cells C2:C96. Copy Cell C96 and paste to Cell C97. The output may be formatted further by changing the number of decimal places in Column C to 2.

Discussing the Outcome

The **Exponential Smoothing** tool puts formulas into the spreadsheet. It uses the first observation as the forecast for the second period. For example, to obtain the forecast for Period 30:

$$\text{Forecast}(30) = 0.2(-14.51) + 0.8(-12.36) = -12.79$$

Likewise, the forecast for Period 97 is −7.38. Some authors suggest using an average of the first n observations.

PROBLEMS

1. Refer to Problem 1 in Section 13.1 (p. 180). Smooth the time series using exponential smoothing with a smoothing constant of 0.1.

2. Refer to Problem 2 in Section 13.1 (p. 180). Smooth the time series using exponential smoothing with a smoothing constant of 0.2.

3. Refer to Problem 3 in Section 13.1 (p. 180). Smooth the time series using exponential smoothing with a smoothing constant of 0.3.

13.3 TREND FITTING

Trend represents the long-run growth or decline in a time series. Factors influencing trend could be technological change, changes in consumer tastes, market growth, increases in income or population, or inflation.

The *linear* trend line is a simple least-squares regression of y on the time variable x_t:

$$T_t = b_0 + b_1 x_t, \text{ where } x_t = 1 \text{ in Period 1, } x_t = 2 \text{ in Period 2, etc.}$$

The first observation in the time series is assigned a value of $x_t = 1$. The remaining observations are then assigned the values 2, 3, 4, etc., in consecutive order.

In order to project the trend into the future, consider the following example:

$$T_t = 12.792 + 0.475 x_t, \text{ where } x_t = 1 \text{ in Quarter 1 of 1994 and } x_t = 20$$
$$\text{in Quarter 4 of 1998}$$

To project the trend to Quarter 3 of 1999, substitute $x_t = 23$ into the above equation.

The quadratic trend model is as follows:

$$T_t = b_0 + b_1 x_t + b_2 x_t^2, \text{ where } x_t = 1 \text{ in Period 1, } x_t = 2 \text{ in Period 2, etc.}$$

The quadratic variable is created by squaring the values of the time variable x_t and would then be used as the second independent variable in the model.

The **Data Analysis: Regression** tool discussed in Section 11.2 can be used to construct both a linear trend model and a quadratic trend model.

PROBLEMS

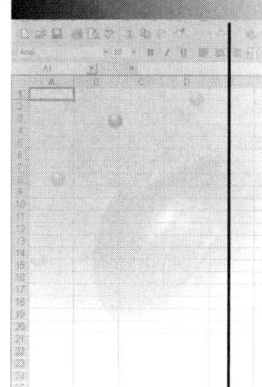

1. Refer to Problem 1 in Section 13.1 (p. 180). Fit a least-squares linear trend equation and a quadratic trend equation to the data. Which is the more appropriate model?

2. Refer to Problem 2 in Section 13.1 (p. 180). Fit a least-squares linear trend equation and a quadratic trend equation to the data. Which is the more appropriate model?

3. Refer to Problem 3 in Section 13.1 (p. 180). Fit a least-squares linear trend equation and a quadratic trend equation to the data. Which is the more appropriate model?

13.4 CLASSICAL MULTIPLICATIVE DECOMPOSITION

The multiplicative time-series model is as follows:

$$y = T \times C \times S \times I$$

where T = trend (long-term growth or decline of the time series)
C = cyclical (alternating periods of expansion and contraction of *more* than one year's duration)
S = seasonal (repetitive patterns completing themselves *within* a year— that is, quarterly or monthly data)
I = irregular (residual movements in the time series, after all other factors have been removed, that follow no recognizable pattern)

The trend component is expressed in the original units of the time series while the other three components are expressed as indexes.

This model is used for increasing or decreasing seasonal variation, data with either no trend or a constant trend, and medium- to long-range forecasting.

The time-series components can have the following causes:

Trend:
- technological change
- changes in consumer tastes and preferences
- changes in per capita income
- changes in population
- inflation or deflation
- market growth

Cyclical:
- business cycles
- weather cycles
- changes in fashion styles

Seasonal:
- weather
- customs

Irregular:
- wildcat strikes
- severe weather conditions (hurricanes, snow storms, etc.)
- accidents
- wars
- earthquakes

Excel does not have a **Data Analysis** tool to perform classical multiplicative decomposition, so Excel functions and formulas can be used instead. The steps for classical multiplicative decomposition, as well as its uses, are shown in the following example.

EXAMPLE — SALES TRENDS

Chap13\Ex13-4.xls

Consider the following time series representing quarterly sales (in $1,000s) for a small eastern Canadian company:

SECTION 13.4 • CLASSICAL MULTIPLICATIVE DECOMPOSITION

EXAMPLE

Year	Quarter	Time Series
1992	1	15
	2	20
	3	8
	4	19
1993	1	16
	2	21
	3	11
	4	17
1994	1	16
	2	25
	3	14
	4	23
1995	1	18
	2	26
	3	12
	4	22
1996	1	19
	2	25
	3	15
	4	26

Determine the seasonal indexes and seasonally adjusted sales figures.

Setting Up the Template

1. Type the headings and data above or import the headings and data from Columns A, B, and C of file **Chap13\Ex13-4.xls**. Enter the headings in Cells D1:E2. Select Cells A1:F1. From the **Format** menu, choose **Column, Width**. Change the width to **13**. To format the headings, select Cells A1: E2 and click **Bold** on the toolbar.

2. In Cell D5, enter the formula **=(C3+C4+C5+C6+C4+C5+C6+C7)/8** to produce the four-quarter centred moving average ($T \times C$) for Quarter 3 of 1992. The objective is to separate the seasonal and irregular variation from the trend and cyclical variation. Copy Cell D5 and paste to Cells D6:D20 to produce the remaining four-quarter centred moving averages.

3. In Cell E5, enter the formula **=C5/D5*100** to produce the ratio to moving average (specific seasonal relative) for Quarter 3 of 1992. Copy Cell E5 and paste to Cells E6:E20 to produce the remaining ratio to moving averages

$[y / (T \times C) = S \times I]$. Change the number of decimal places in Cells D5:E20 to **3**.

4. Enter the headings in Cells B28:F28 and Cells A29:A36. To format, select the headings and click **Bold** on the toolbar. Copy the ratio to moving average for Quarter 3 of 1992 by entering **=E5** in Cell D29. Arrange the remaining ratio to moving average values according to their respective seasons by using the appropriate cell references (that is, enter **=E6** in Cell E29, **=E7** in Cell B30, **=E8** in Cell C30, etc.).

5. In Cell B35, enter the function **=MEDIAN(B29:B33)** to compute the median ratio to moving average for Quarter 1. Copy Cell B35 and paste to Cells C35:E35 to compute the medians for Quarters 2 to 4.

6. In Cell F35, enter the function **=SUM(B35:E35)** to compute the sum of the medians for the four quarters.

7. In Cell B36, enter the formula **=(B35*400)/$F$35** to compute the seasonal index for Quarter 1. This is done by multiplying the median for Quarter 1 by the adjustment factor [adjustment factor = (number of seasons × 100)/sum of medians]. Copy Cell B36 and paste to Cells C36:E36 to compute the seasonal indexes for Quarters 2 to 4.

8. In Cell F36, enter the function **=SUM(B36:E36)** to verify that the sum of the seasonal indexes is equal to 400. Change the number of decimal places in Cells B35:F36 to **3**.

9. Enter the headings in Cells A49:E50. To format, select the headings and click **Bold** on the toolbar. Enter or copy the data into Cells A51:C70.

10. In Cells D51:D54, enter **=$B$36**, **=$C$36**, **=$D$36**, and **=$E$36**, respectively, to copy the seasonal indexes. Copy Cells D51:D54 and paste to Cells D55:D58, D59:D62, D63:D66, and D67:D70.

11. In Cell E51, enter the formula **=C51/D51*100** to compute the seasonally adjusted figure for Quarter 1 of 1992. Copy Cell E51 and paste to Cells E52:E70 to compute the remaining seasonally adjusted figures. Change the number of decimal places in Cells E51:E70 to **3**.

12. A chart depicting the actual time series versus the deseasonalized time series can be produced using a line chart as discussed in Chapter 1. (To create a chart using nonadjacent columns, in Step 2 of the **Chart Wizard** highlight one column then, holding down CTRL+ALT, highlight the other.)

Discussing the Outcome

The seasonal pattern indicates that sales are highest in Quarter 2 and then fall to a seasonal low in Quarter 3. For monthly data, 12-month centred moving averages would be taken and the seasonal indexes would sum to 1200; otherwise the derivation of the seasonal indexes is the same.

Uses of Seasonal Indexes

1. **Deseasonalization:** Given the actual time series data y, divide by the seasonal index S (in decimal form) to isolate the seasonally adjusted figure ($y/S = T \times C \times I$). Here the seasonal influence on the data is removed:

   Seasonally adjusted figure for Quarter 4 of 1994 = $23/113.002 \times 100 = 20.354$

   It is evident from the chart on the spreadsheet that the deseasonalized data are smoother and do not have the peaks and troughs; therefore, they are more reflective of the long-term movement of the time series.

2. **Forecasting:** Given $T \times C$, multiply by the seasonal index S (in decimal form) to obtain a more realistic forecast. The trend line is a simple least-squares regression of y on the time variable x_t:

   $T_t = 14.579 + 0.364x_t$, where $x_t = 1$ in Quarter 1 of 1992 and $x_t = 20$ in Quarter 4 of 1996

   Given $C = 100$, forecast y for Quarter 2 of 1998:

   $$\hat{y} = [14.579 + 0.364(26)] \times (129.658/100) = 31.174$$

Note that variations of the model may fit the trend line through the seasonally adjusted series or obtain the seasonal indexes from the detrended series.

Classical Decomposition Problems and Limitations

1. There are no trading day adjustments.
2. Values are lost at the beginning and end.
3. This is a simplistic approach to obtaining seasonal indexes.
4. Seasonal indexes are constant over time.

PROBLEMS

1. Refer to Problem 2 in Section 13.1 (p. 180).

    a. Compute the four seasonal indexes.
    b. Obtain the deseasonalized value for Quarter 2 of 1992.
    c. Obtain a forecast for Quarter 3 of 1999, including linear trend and seasonal effects only.
    d. Determine the combined cyclical and irregular components for Quarter 4 of 1994.

2. Refer to Problem 3 in Section 13.1 (p. 180).

    a. Plot the time series. Does a seasonal component seem to be present? Describe the seasonal pattern in the time series.
    b. Obtain the quarterly seasonal indexes for the time series, assuming the seasonal pattern is stable.

    Chap13\P13-4#3.xls

3. The following data represent the monthly sales of a product for a New Brunswick manufacturer from 1991 to 1995:

	1991	1992	1993	1994	1995
January	740	742	895	952	1034
February	699	701	792	862	1033
March	775	770	884	937	1127
April	897	931	1057	1108	1286
May	1031	1093	1205	1275	1465
June	1106	1228	1327	1425	1636
July	1165	1293	1308	1486	1619
August	1214	1348	1434	1553	1604
September	1209	1349	1479	1606	1523
October	1130	1297	1454	1602	1422
November	972	1067	1171	1401	1110
December	785	903	1022	1208	1018

   a. Obtain the deseasonalized value for July 1992.
   b. Prepare forecasts for the first 6 months of 1996, assuming that the cycle will be equal to 103 in each of those months.
   c. Determine the combined cyclical and irregular components for October 1994.

13.5 REGRESSION APPROACH TO TIME SERIES

An *additive model* should be used for data exhibiting constant seasonal variation, data with either no trend or constant trend, and for medium- to long-range forecasting.

We will assume a *linear* trend and a length of seasonality of four ($L = 4$). The linear trend component may be described as in Section 13.3, and the seasonal components may be described using indicator variables as follows:

$$y_t = \beta_0 + \beta_1 x_t + \beta_2 z_1 + \beta_3 z_2 + \beta_4 z_3 + \varepsilon_t$$

where $x_t = 1$ in Period 1, $x_t = 2$ in Period 2, etc.
$z_1 = 1$ if Quarter 1, $z_1 = 0$ if Quarter 2 or 3, $z_1 = -1$ if Quarter 4
$z_2 = 1$ if Quarter 2, $z_2 = 0$ if Quarter 1 or 3, $z_2 = -1$ if Quarter 4
$z_3 = 1$ if Quarter 3, $z_3 = 0$ if Quarter 1 or 2, $z_3 = -1$ if Quarter 4

To represent four possible conditions, we must create three indicator variables. (If monthly data were used, 11 indicator variables would have to be used.) The seasonal indexes therefore become

$$S_1 = 100 + b_2$$
$$S_2 = 100 + b_3$$
$$S_3 = 100 + b_4$$
$$S_4 = 100 - b_2 - b_3 - b_4 = 400 - S_1 - S_2 - S_3$$

The **Data Analysis: Regression** tool discussed in Section 12.1 can be used to construct such a multiple regression model. The steps for constructing a regression model using a linear trend component and quarterly indicator variables are shown in the following example.

EXAMPLE

SALES FORECAST

Chap13\Ex13-5.xls

Consider the following time series representing quarterly sales (in $1,000s) for a small Canadian company:

15, 22, 10, 17, 14, 21, 11, 19, 16, 25, 14, 21, 18, 25, 12, 20, 17, 26, 15, 24

Construct a regression model using a linear trend component and quarterly indicator variables. Make a forecast of sales for Period 22.

Setting Up the Template

1. Import the headings and data from Columns A to E of file **Chap13\Ex13-5.xls**.

2. From the **Tools** menu, choose **Data Analysis**. From the **Data Analysis** tools list, choose **Regression** and click **OK**.

3. In the **Regression** dialog box, enter **A1:A21** for the **Input Y Range** and enter **B1:E21** for the **Input X Range**. Check the **Labels** box. Choose the **Output Range** option button, then click in the box and enter **G1** for the **Output Range**. Click **OK**.

4. The results will be produced, but notice that some of the titles do not fit in the columns. Rather than individually formatting each column's width, select all the output range and, from the **Format** menu, choose **Column, AutoFit Selection**. The duplicate "Lower 95.0%" and "Upper 95.0%" at the far right of the output can be deleted. The output may be formatted further by changing the number of decimal places of numerical results or by italicizing headings, as you would format parts of any other spreadsheet.

Discussing the Outcome

The coefficients for the estimated regression line can be found in Cells H17:H21 of the third section of the summary output. The equation for the estimated regression line is therefore

$$\hat{y}_t = 15.08125 + 0.2875 x_t - 1.66875 z_1 + 5.84375 z_2 - 5.84375 z_3$$

The coefficient of determination in Cell H5 indicates that approximately 95.5% of the variation in sales is explained by the linear trend variable and quarterly indicator variables.

The seasonal indexes would therefore become

$$S_1 = 100 - 1.66875 = 98.33125$$
$$S_2 = 100 + 5.84375 = 105.84375$$
$$S_3 = 100 - 5.84375 = 94.15625$$
$$\begin{aligned} S_4 &= 100 + 1.66875 - 5.84375 + 5.84375 \\ &= 400 - 98.33125 - 105.84375 - 94.15625 \\ &= 101.66875 \end{aligned}$$

The seasonal pattern indicates that sales are highest in Quarter 2 and then fall to a seasonal low in Quarter 3. Since the *p*-values for the four variables in Cells K18:K21 are all less than a 0.05 level of significance, it can be concluded that all four independent variables (the linear trend variable and the three seasonal variables) are significant and therefore should be kept in the model.

To forecast for Period 22:

$$\hat{y} = 15.08125 + 0.2875(22) + 5.84375(1) = 27.25$$

PROBLEMS

1. Refer to Problem 2 in Section 13.1 (p. 180). Construct a regression model using a linear trend component and quarterly indicator variables. Make forecasts for the next eight quarters.

2. Refer to Problem 3 in Section 13.1 (p. 180). Construct a regression model using a linear trend component and quarterly indicator variables. Make forecasts for the next four quarters.

3. Refer to Problem 3 in Section 13.1 (p. 180). Construct a regression model using a quadratic trend component and quarterly indicator variables. Make forecasts for the next four quarters.

CHAPTER 14
Quality Control Charts

14.1 $\bar{x}$ AND R CHARTS

A *control chart* gives a chronological view of a process by plotting data based on samples selected from the process, and it indicates whether or not a process is *in control*. A process is in control if the variability is *random* and due to *common* causes rather than assignable (special) factors. A process is *out of control* when one or more observations fall outside the control limits or when all the observations fall within the control limits but an unusual pattern is detected.

A control chart will include a centre line, upper and lower control limits, and a set of points representing a sample of items from a process. Any point outside the control limits should be analysed to determine what cause is responsible for this extreme point. Even if a process is in control, unusual patterns may indicate that there is a shift in a process or that the process is losing control. Patterns would not exist if only random variability were present.

$\bar{x}$ and R charts are used with variable data (that is, distance, width, or temperature) to determine quickly whether there has been a shift in the process mean or process variability over time. If a shift is detected, it indicates that the process is out of control and requires some sort of remedial action. The values of $\bar{x}$ or R are plotted on the y-axis against the sequence of samples over time on the x-axis.

R CHARTS

An R chart is used to monitor the *variability* of a process. It does this by selecting a sample of observations at specified time intervals and monitoring the sequential sample ranges.

Centre line: $\bar{R}$ = mean of the ranges from k repeated samples of size n

$$\text{Upper control limit (UCL): } D_4 \bar{R}$$
$$\text{Lower control limit (LCL): } D_3 \bar{R}$$

where D_3 and D_4 are taken from a table of control chart factors

A process is deemed to be out of control at period j if $R_j <$ LCL or if $R_j >$ UCL.

$\bar{x}$ CHARTS

An $\bar{x}$ chart is used to monitor the *mean* of a process. It does this by selecting a sample of observations at specified time intervals and monitoring the sequential sample means.

Centre line: $\bar{\bar{x}}$ = mean of the sample means from k repeated samples of size n

$$\text{Upper control limit (UCL): } \bar{\bar{x}} + A_2 \bar{R}$$
$$\text{Lower control limit (LCL): } \bar{\bar{x}} - A_2 \bar{R}$$

where A_2 is taken from a table of control chart factors

A process is deemed to be out of control at period j if $\bar{x}_j <$ LCL or if $\bar{x}_j >$ UCL.

Note that the upper and lower control limits for the $\bar{x}$ chart are calculated using $\bar{R}$; thus, before using an $\bar{x}$ chart, we must be certain that the variability is in control by constructing the R chart first. Excel does not have a **Data Analysis** tool to calculate the centre line and upper and lower control limits the two control charts, so Excel functions and formulas can be used instead. The steps for setting up the control charts are shown in the following example.

EXAMPLE

INTERNET SERVICE CALLS

Chap14\Ex14-1.xls

An Internet service provider in Montreal is interested in monitoring the length of customer service calls. Every day the company randomly samples 5 calls received that day. For each call, the length is recorded in minutes:

Day	Sample #1	Sample #2	Sample #3	Sample #4	Sample #5	Mean	Range
1	17	22	11	19	14	16.6	11
2	25	21	20	15	18	19.8	10
3	19	16	25	10	15	17.0	15
4	17	25	14	25	21	20.4	11
5	25	17	25	17	16	20.0	9
6	17	20	19	22	17	19.0	5
7	20	11	22	11	25	17.8	14
8	13	25	19	24	11	18.4	14
9	22	21	23	11	16	18.6	12
10	15	13	19	24	19	18.0	11
11	16	13	18	14	25	17.2	12
12	17	19	14	19	20	17.8	6
13	18	18	12	25	24	19.4	13
14	19	19	19	24	13	18.8	11
15	15	23	19	22	15	18.8	8
16	10	15	16	15	22	15.6	12
17	15	14	18	24	16	17.4	10
18	12	14	18	13	16	14.6	6
19	21	12	11	19	15	15.6	10
20	13	23	24	18	18	19.2	11

a. Calculate the centre line and control limits for an R chart.
b. Calculate the centre line and control limits for an $\bar{x}$ chart.
c. Does the process appear to be in control? Explain.

Setting Up the Template

1. Type or import the headings in Cells A1:A7, D1:D5, and A10:J10 of file **Chap14\Ex14-1.xls**. Select Cells A1:J1. From the **Format** menu, choose **Column, Width**. Change the width to **12**. To format the headings, select them and click **Bold** on the toolbar.

2. Import the data from Cells A11:D30 of the file, representing day number, sample size, sample mean, and sample range. If *only* the raw data were given (Sample #1 through Sample #5 in the table above), the **AVERAGE**, **MAXIMUM**, and **MINIMUM** functions could be used on the data to compute the sample mean and sample range. To format, select Cells A10:J30 and click **Center** on the toolbar.

EXAMPLE

3. In Cell B1, enter the function **=AVERAGE(D11:D30)** to compute the value of $\bar{R}$.

4. In Cell B2, enter the function **=AVERAGE(B11:B30)** to compute the value of $\bar{n}$.

5. In Cells B3 and B4, enter the values of D_3 and D_4, respectively. These values come from a statistical table of control chart factors. Using our example, enter **0** in Cell B3 and **2.114** in Cell B4.

6. In Cell B5, enter the formula **=B3*B1** to compute the LCL for the R chart.

7. In Cell B6, enter **=B1** to duplicate the centre line for the R chart.

8. In Cell B7, enter the formula **=B4*B1** to compute the UCL for the R chart.

9. In Cell E1, enter the function **=AVERAGE(C11:C30)** to compute the value of $\bar{\bar{x}}$.

10. In Cell E2, enter the value of A_2 (**0.577**), which comes from a table of control chart factors.

11. In Cell E3, enter the formula **=E1-E2*B1** to compute the LCL for the $\bar{x}$ chart.

12. In Cell E4, enter **=E1** to duplicate the centre line for the $\bar{x}$ chart.

13. In Cell E5, enter the formula **=E1+E2*B1** to compute the UCL for the $\bar{x}$ chart.

14. Once the control limits and centre lines have been computed for the R chart and $\bar{x}$ chart, we need to copy them to columns E through J so that the **Chart Wizard** can then be used to draw the respective control charts. Enter **=$B$5, =$B$6, =$B$7, =$E$3, =$E$4, =$E$5** in Cells E11:J11, respectively. Copy Cells E11:J11 and paste to E12:J30.

15. Use the **Chart Wizard** to produce *xy* scatter charts as described in Section 1.4 (pp. 18–20). Select the **Chart sub-type** that shows points connected with straight lines. In Step 2, use **Data Range: A10:A30,D10:G30** for the R chart, and **Data Range: A10:A30,C10:C30,H10:J30** for the $\bar{x}$ chart. Label the *x*-axis **Day** and the *y*-axis **Minutes**. To make a control chart more visually appealing, double-click on the UCL line. From the **Format Data Series** dialog box, select the **Patterns** tab. Under **Marker**, select the **None** option button. The UCL line changes to a smoothed line. Repeat the same procedure for the LCL line and the centre line. The colours of the lines can also be changed as desired.

EXAMPLE

Discussing the Outcome

Both charts seem to be in statistical control, as all of the ranges and sample means fall within their respective control limits for all 20 days.

Now that the template has been created, we can observe the effect of changes in the data on either the R chart or the $\bar{x}$ chart. For example, change the value of the sample mean for Day 4 in Cell C14 from 20.4 to 26.4. Notice how the mean LCL becomes 12.21265, the centre line becomes 18.3, and the mean UCL becomes 24.38735. The $\bar{x}$ chart changes correspondingly to reflect these values. Now observe that the mean for Day 4 is above the UCL and that the process is deemed to be out of control at Day 4.

PROBLEMS

1. The following is a list of sample means and ranges:

$\bar{x}$	2.3	3.1	4.8	2.6	3.9	3.0	3.5	2.7
R	0.5	1.1	0.8	1.2	0.3	0.6	1.0	0.4

Assuming that each sample size is 15, construct the $\bar{x}$ chart. Does the process appear to be in control? Explain.

2. The following is a list of sample means and ranges:

$\bar{x}$	2.0	3.2	4.6	2.9	3.5	2.4	3.6	2.9
R	0.7	1.0	0.9	1.3	0.5	0.8	1.1	0.6

Assuming that each sample size is 12, construct the R chart. Does the process appear to be in control? Explain.

3. The following is a list of sample means and ranges:

$\bar{x}$	2.1	3.3	4.5	2.7	3.2	3.2	3.9	2.6
R	0.5	1.1	0.8	1.2	0.3	0.6	1.0	0.4

Assuming that each sample size is 16, construct the $\bar{x}$ chart. Does the process appear to be in control? Explain.

4. The following is a list of sample means and ranges:

$\bar{x}$	2.1	3.0	4.4	2.6	3.4	2.1	3.5	2.7
R	0.9	1.1	1.2	1.5	0.7	0.5	1.3	0.7

Assuming that each sample size is 10, construct the R chart. Does the process appear to be in control? Explain.

14.2 p AND c CHARTS

These control charts are to be used with attribute data—data with two possible outcomes (that is, good/defective).

p CHARTS

A p chart is used to monitor the proportion of defective items for each sample.

Centre line: $\bar{p}$ = proportion of defective items over k samples
= (total defectives)/(total observations)

Upper control limit (UCL): $\bar{p} + 3 \sqrt{\dfrac{\bar{p}(1-\bar{p})}{n}}$

Lower control limit (LCL): $\bar{p} - 3 \sqrt{\dfrac{\bar{p}(1-\bar{p})}{n}}$

If the LCL is negative, set it equal to 0.

c CHARTS

A c chart is used to monitor the number of defective items for each sample.

Centre line: $\bar{c}$ = mean number of defective items over k samples

Upper control limit (UCL): $\bar{c} + 3\sqrt{\bar{c}}$
Lower control limit (LCL): $\bar{c} - 3\sqrt{\bar{c}}$

If the LCL is negative, set it equal to 0.

For both the p chart and the c chart, a process is said to be out of control when data points exceed the control limits, show evidence of a pattern, or are improperly disbursed on the chart.

Excel does not have a **Data Analysis** tool to calculate the centre line and the upper and lower control limits for the two control charts, so Excel functions and formulas can be used instead. The steps for setting up the control charts are shown in the following example.

EXAMPLE: SCANNING BAR CODES

Chap14\Ex14-2.xls

Checkout lines in Canadian grocery stores use price scanners to record the price of a product, by running the item's bar code over the scanner. If the scanner fails to read a bar code, then the cashier has to type the code or price into the register manually. The rejection rate is of course affected by the quality of the bar codes on the product packages. In order to monitor the quality of the bar codes, a grocery store randomly samples 1000 products a day for 30 days to determine the number of products rejected by the scanners. The results are shown in the following table:

Day	Rejected
1	5
2	9
3	13
4	7
5	14
6	5
7	16
8	9
9	14
10	7
11	9
12	14
13	6
14	12
15	10
⋮	⋮

(complete data on disk)

a. Calculate the centre line and the control limits for a p chart.
b. Calculate the centre line and the control limits for a c chart.
c. Does the process appear to be in control? Explain.

Setting Up the Template

EXAMPLE

1. Type or import the headings in Cells A1:A5, D1:D4, and A8:J8 of file **Chap14\Ex14-2.xls**. To format the headings, select them and click **Bold** on the toolbar. Select Cells A1:J1. From the **Format** menu, choose **Column, Width**. Change the width to **12**.

2. Import the data from Cells A9:C38 of the file, representing day number, sample size, and the number of defective items for each sample. To format, select Cells A8:J38 and click **Center** on the toolbar.

3. In Cell B1, enter the function **=AVERAGE(B9:B38)** to compute the value of $\bar{n}$.

4. In Cell D9, enter the formula **=C9/B9** to compute the proportion of defective items for the first sample. Copy Cell D9 and paste to Cells D10:D38.

5. In Cell B2, enter the function **=AVERAGE(D9:D38)** to compute the value of $\bar{p}$.

6. In Cell B3, enter the formula **=B2-3*SQRT(B2*(1-B2)/B1)** to compute the LCL for the p chart.

7. In Cell B4, enter **=B2** to duplicate the centre line for the p chart.

8. In Cell B5, enter the formula **=B2+3*SQRT(B2*(1-B2)/B1)** to compute the UCL for the p chart.

9. In Cell E1, enter the function **=AVERAGE(C9:C38)** to compute the value of $\bar{c}$.

10. In Cell E2, enter the formula **=E1-3*SQRT(E1)** to compute the LCL for the c chart.

11. In Cell E3, enter **=E1** to duplicate the centre line for the c chart.

12. In Cell E4, enter the formula **=E1+3*SQRT(E1)** to compute the UCL for the c chart.

13. Once the control limits and centre lines have been computed for the p chart and c chart, we need to copy them to Columns E to J so that the **Chart Wizard** can then be used to draw the respective control charts. Enter **=$B$3, =$B$4, =$B$5, =$E$2, =$E$3,** and **=$E$4** in Cells E9:J9, respectively. Copy Cells E9:J9 and paste to E10:J38.

EXAMPLE

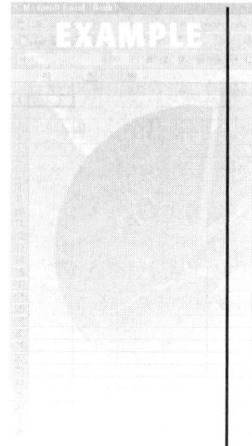

14. Use the **Chart Wizard** to produce *xy* scatter charts as described in Section 1.4 (pp. 18–20). Select the **Chart sub-type** that shows points connected with straight lines. In Step 2, use **Data Range: A8:A38,D8:G38** for the *p* chart and **Data Range: A8:A38,C8:C38,H8:J38** for the *c* chart. Label the *x*-axis **Day**. Label the *y*-axis **Proportion** for the *p* chart and **Rejected** for the *c* chart. To make the control chart more visually appealing, double-click on the UCL line. From the **Format Data Series** dialog box, select the **Patterns** tab. Under **Marker**, select the **None** option button. The UCL line changes to a smoothed line. Repeat the same procedure for the LCL line and the centre line. The colours of the lines can also be changed as desired.

Discussing the Outcome

Both charts seem to be in statistical control, as all of the values of *p* and c fall within their respective control limits for all 30 days.

Now that the template has been created, we can observe the effect of changes in the data on either the *p* chart or the *c* chart. For example, change the number of products rejected by the scanners on Day 25 in Cell C33 from 16 to 21. Notice how the LCL for the *p* chart becomes 0.000704, the centre line becomes 0.010267, and the UCL becomes 0.019830. The *p* chart changes correspondingly to reflect these values. Now observe that the proportion for Day 25 is above the UCL and that the process is deemed to be out of control at Day 25.

PROBLEMS

Chap14\P14-2#1.xls

1. A manufacturer of a particular product took 9 samples of size 60. The number of defectives in each of these samples is as follows:

Sample	Number of Defectives
1	10
2	11
3	8
4	15
5	5
6	14
7	22
8	11
9	12

Construct the *p* chart. Does the process appear to be in control? Explain.

PROBLEMS

Chap14\P14-2#2.xls

2. A computer manufacturer of a particular component took 7 samples of size 60. The number of defects in each of these samples is as follows:

Sample	Number of Defectives
1	18
2	21
3	22
4	26
5	21
6	14
7	15

Construct the c chart. Does the process appear to be in control? Explain.

Chap14\P14-2#3.xls

3. A manufacturer of a television transistor took 8 samples of size 50. The number of defects in each of these 8 samples is as follows:

Sample	Number of Defectives
1	11
2	12
3	10
4	7
5	24
6	20
7	6
8	10

Construct the p chart. Does the process appear to be in control? Explain.

Chap14\P14-2#4.xls

4. A manufacturer of a television transistor took 10 samples of size 50. The number of defects in each of these samples is as follows:

Sample	Number of Defectives
1	13
2	11
3	14
4	8
5	23
6	21
7	7
8	12
9	11
10	13

Construct the c chart. Does the process appear to be in control? Explain.

Index

ABS function, 65, 94, 98
Add Trendline command, 146
Add-Ins, viii, xii, 31
additive model, 191
adjusted R-square, 165
analysis of variance (ANOVA), 138
Analysis ToolPak, xii, 31
Anova: Single Factor, Data Analysis tool, 140
ANOVA table, 138
ANOVA table for regression, 165
assignable factors, 194
attribute data, 199
AVERAGE function, xii, 29, 74, 83, 159, 197, 201

bar chart, 7
bin range, 38
BINOMDIST function, xii, 54
binomial distribution, 52

c chart, 199
categorical variables, 42, 45
cell addresses, xi
central limit theorem, 73, 77, 93
centre line, 194, 199
Chart Wizard, 2, 8, 13, 18, 146, 188, 197, 202
CHIDIST function, 130
CHIINV function, 130, 135
chi-square test for normality, 128
CHITEST function, 135
classical multiplicative decomposition, 185
coefficient of correlation, 151, 168
coefficient of determination, 151
coefficient of multiple determination, 165
coefficient of variation, 34
column chart, 7
confidence interval for difference between two means, 106, 112

confidence interval for mean of dependent variable, 157
confidence interval for population mean, 77, 81
confidence interval for population proportion, 85
confidence interval for regression coefficients, 152, 164
contingency table test for independence, 132
control chart, 194
control chart factors, 195
control limits, 194, 199
copying, ix
correlation, 21, 168
Correlation, Data Analysis tool, 170
COUNT function, 29, 83
cumulative relative frequency polygon, 37
cyclical component, 185

Data Analysis tools, xii. *See also names of individual tools*
degrees of freedom for chi-square distribution, 128, 133
degrees of freedom for F-distribution, 123, 139
degrees of freedom for t-distribution, 82, 97, 105
dependent variable, 17, 144
descriptive statistics, 28, 33
Descriptive Statistics, Data Analysis tool, 31, 78, 82
deseasonalization, 189
differences, 112

editing charts, 4, 10, 21, 39
error term, 144, 154, 161
expected value, 49
EXPONDIST function, 68
exponential distribution, 67
exponential smoothing, 181
Exponential Smoothing, Data Analysis tool, 182
extrapolation, 145, 150

F-distribution, 123
F-test for significance of regression line, 165
F-test for two population variances, 123
F-Test Two-Sample for Variances, Data Analysis tool, 125
finite population correction factor (FPCF), 57
forecast, 181, 183, 189
Format, Cells command, 2, 109
Format, Column, AutoFit Selection command, 32, 108, 115, 126, 141, 150, 156, 163, 171, 174, 178, 192

INDEX

Format, Column, Width command, 2
formatting cells, ix
formatting charts, 4, 10, 21, 39
formula bar, vii, x
formulas, x
frequency distribution, 25
FREQUENCY function, 26
frequency polygon, 15
functions, xi. *See also names of individual functions*

histogram, 37
Histogram, Data Analysis tool, 38, 74
hypergeometric distribution, 56
HYPGEOMDIST function, 58
hypothesis test for difference in two proportions, 119
hypothesis test for population mean, 93, 97
hypothesis test for population proportion, 101
hypothesis test for two population means, 105, 112
hypothesis test for two population variances, 123

IF function, 94, 95, 98, 99, 102, 103, 121, 130, 135
importing data, viii
independent variable, 17, 144
indicator variable, 172, 191
interaction, 175
interquartile range, 35
interval estimate, 77
irregular component, 185

KURT function, 29
kurtosis, 33

least squares regression line, 145
line chart, 11
linear relationship, 152, 164, 168
linear trend, 184, 189

matched samples, 112
mathematical expectation, 49
MAX function, 29
maximum, 33
mean, 33, 53, 57, 60, 63, 67
mean of a process, 195

median, 33
MEDIAN function, 29, 188
menu bar, vii
MIN function, 29
minimum, 33
mode, 33
MODE function, 29
moving average, 177
Moving Average, Data Analysis tool, 178
multicollinearity, 169
multiple regression, 161
multiplicative time-series model, 185

normal distribution, 63
NORMDIST function, 65, 129, 130
NORMINV function, 65
NORMSDIST function, 94, 95, 102, 103, 121
NORMSINV function, 78, 86, 88, 91, 94, 95, 102, 103, 120, 121

one-way summary table for categorical variables, 42
opening files, viii
ordered array, 24

p chart, 199
p-value, 93, 97, 101, 106, 113, 119, 124, 128, 133, 152, 160, 164
page setup, xiii
parameter, 144, 161
Paste Function, xii
pasting, ix
percentile, 35
PERCENTILE function, 29, 30
pie chart, 1
pilot sample, 87, 90
PivotTable Wizard, 43, 46, 85, 119, 129, 133
pivot table, 42
planning value, 87, 90
point estimate, 77
Poisson distribution, 59
POISSON function, 61
pooled variance, 106, 119
population mean, 77, 81, 87, 93, 97, 105, 112
population proportion, 85, 90, 101, 119

population variance, 87, 123
prediction, 145
prediction interval, 158
predictor variable, 17, 162
printing, xiii
probability distribution, 49

quadratic regression, 175
quadratic trend, 184
quality control, 194

R chart, 194
RANDBETWEEN function, 71
Random Number Generation, Data Analysis tool, 74
random sample, 70
random variable, 49
range, 33
regression coefficient, 150, 161, 163, 169
Regression, Data Analysis tool, 149, 155, 162, 173, 192
regression line, 144, 154, 161
relative frequency polygon, 15
residual, 154, 157
residual plot, 154
ROUNDUP function, 88, 91

sample size for population mean, 87
sample size for population proportion, 90
sampling, 70
Sampling, Data Analysis tool, 71
sampling distribution, 73, 75
saving files, viii
scatter chart, 17, 145, 154, 197, 202
seasonal component, 185
simple linear regression, 144, 148
SKEW function, 29
skewness, 33
slope, 144
smoothing constant, 181
Sort command, 25
sorting data, 24
SQRT function, 50, 54, 58, 78, 79, 83, 86, 94, 98, 102, 120, 159, 201
standard deviation, 34, 49, 63, 67

standard error, 34
standard error of the estimate, 153, 159, 166
STDEV function, 29, 74, 83
SUM function, 29, 50, 130, 134, 135, 159, 188
sum of squares, 138
summary tables for categorical variables, 42, 45
SUMPRODUCT function, 50

t-distribution, 81, 97, 105
t-test for one population mean, 97
t-test for two population means, 105, 112
t-Test: Paired Two Sample for Means, Data Analysis tool, 114
t-Test: Two Sample Assuming Equal Variances, Data Analysis tool, 107
t-tests for independent variables, 152, 164, 175, 193
TDIST function, 98
Text Import Wizard, viii
time series, 177, 181, 184, 185, 191
time-series regression, 191
TINV function, 83, 98, 99, 109, 115, 159
toolbars, vii
treatment, 138
trend component, 185
TREND function, 147, 159
trend line, 144, 184
two-way summary table for categorical variables, 45

VAR function, 29
variability of a process, 194
variance, 34, 49, 53, 57, 60

worksheets, vii

$\bar{x}$ chart, 194

y-intercept, 144

z-distribution, 77
z-test for differences in two proportions, 119
z-test for one population mean, 93